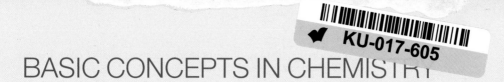

Stereochemistry

DAVID G. MORRIS

University of Glasgow

**WILEY-
INTERSCIENCE**

RS•C
ROYAL SOCIETY OF CHEMISTRY

Cover images © Murray Robertson/visual elements 1998–99, taken from the
109 Visual Elements Periodic Table

For ordering and customer service, call 1-800-CALL-WILEY.

Library of Congress Cataloging-in-Publication Data:
Library of Congress Cataloging-in-Publication Data is available.
ISBN: 0-471-22477-4

Typeset in Great Britain by Wyvern 21, Bristol
Printed and bound by Polestar Wheatons Ltd, Exeter

10 9 8 7 6 5 4 3 2 1

Preface

Stereochemistry, first recognized by van't Hoff and Le Bel, now permeates organic chemistry. For a full appreciation of the subject, an understanding of stereochemistry is necessary in terms of both the relevant conventions and definitions that are in use, and also what is happening at a molecular level during a reaction. It is hoped that this self-contained text will provide a means by which undergraduate students (and maybe others) become conversant with stereochemistry. The text being limited in size, it has been necessary to be selective in the choice of topics.

Particular thanks are due to one of my former teachers, Professor Alwyn G. Davies FRS, than whom no-one could have been more helpful and courteous, for reading the manuscript and for his comments. Thanks are likewise due to Mr Martyn Berry. I am grateful to Professor Keiji Gamoh, Dr Sean Higgins, Dr Susan Armstrong and Dr David Rycroft, who read selected chapters and for discussions. I also thank Professor Zvi Rappoport and Dr David Procter for a discussion. The diligence of all the above prevented some errors and infelicities from featuring in the text. For any that remain, I alone am culpable. Gratitude is due to Dr Colin Drayton for help with editing and to Mrs Janet Freshwater of the Royal Society of Chemistry for overseeing the project with benign efficiency.

Space considerations meant that some material had to be omitted. Accordingly, a Glossary of useful stereochemical terms and an addendum to Chapter 7 have been placed on the web at (http://www.wiley.com/go/wiley-rsc). A further such list that derives from IUPAC may be found at the web site address of Dr G. P. Moss (g.p.moss@qmw.ac.uk).

Finally, I thank my wife Brenda for her assistance with the preparation of the manuscript, during which she sacrificed several months of her life on the altar of Stereochemistry.

David G. Morris
Glasgow

BASIC CONCEPTS IN CHEMISTRY

EDITOR-IN-CHIEF	EXECUTIVE EDITORS	EDUCATIONAL CONSULTANT
Professor E W Abel	*Professor A G Davies*	*Mr M Berry*
	Professor D Phillips	
	Professor J D Woollins	

This series of books consists of short, single-topic or modular texts, concentrating on the fundamental areas of chemistry taught in undergraduate science courses. Each book provides a concise account of the basic principles underlying a given subject, embodying an independent-learning philosophy and including worked examples. The one topic, one book approach ensures that the series is adaptable to chemistry courses across a variety of institutions.

TITLES IN THE SERIES

Stereochemistry *D G Morris*
Reactions and Characterization of Solids
  S E Dann
Main Group Chemistry *W Henderson*
d- and f-Block Chemistry *C J Jones*
Structure and Bonding *J Barrett*
Functional Group Chemistry *J R Hanson*

Further information about this series is available at www.wiley.com/go/wiley-rsc

Contents

1
Simple Molecules: Hybridization, Conformation and Configuration

Aims

By the end of this chapter you should be familiar with:

- The representation of three-dimensional molecules in two dimensions
- The hybridization of carbon in alkanes, alkenes and alkynes
- The energy profile associated with conformational changes in ethane as the carbon–carbon bond rotates about its axis
- Eclipsed, staggered, *gauche*, *cis* and *trans* conformations as applied to, for example, propane and butane
- Chair and boat conformations, and ring inversion of cyclohexane
- 1,3-Diaxial interactions in substituted cyclohexanes

1.1   Introduction

Chemistry, like everyday life, takes place overwhelmingly in three dimensions. Stereochemistry embraces the spatial aspects of chemistry and can be considered in two parts. The first deals with the shapes and properties of mainly three-dimensional molecules and involves a knowledge of the terms conformation, configuration and chirality, which are introduced in the first two chapters. The second aspect deals with reactivity and includes the preferred or obligatory direction of approach of reagents, and also the consequences for the nature of the products. In respect of reactivity, it can be said that except for spherical reactants, *e.g.* H^+ and Cl^-, there is almost always a preferred direction of approach of one molecule or ion toward another.

We consider first the simplest organic molecule methane, CH_4, in particular its geometry, its representation in two dimensions and its make-up from the component atoms. In methane, all four carbon-- hydrogen bonds are equivalent and are directed toward the corners of a regular tetrahedron with carbon at the centre. The HCH bond angles are all 109°28′, a value that is found only in molecules with 'tetrahedral' symmetry. The normal depiction of methane is given in **1** and the convention, which is applied generally, for representing this three-dimensional molecule in two dimensions is now described. The central carbon is taken to lie in the plane of the paper, together with any two hydrogens. Most commonly these are the upper and left hydrogens and, as shown in **1**, bonds between each of these two hydrogens and carbon are represented by lines of 'normal' thickness. The thick 'wedge' represents a bond between carbon and the hydrogen in front of the page, and the dashed line indicates a bond between carbon and the hydrogen behind the page.

Note that any three atoms must lie in a plane. One can always draw a plane through three points; in the everyday world, for example, a three-legged stool never rocks. The three atoms that lie in the plane of the paper in **1** are chosen for convenience.

The representation of methane in two dimensions is given.

$$\underset{\textbf{1}}{\overset{\displaystyle H}{\underset{\displaystyle H}{H-C_{\prime\prime\prime\prime\prime}H}}}$$

1.2   Hybridization: Methane

When a bond is formed between, say, C and H, the electronic distribution in the orbitals is perturbed and two atomic orbitals that normally each contain an electron are said to overlap. In the case of a simple molecule such as in methane, the question arises: how does one reconcile the equivalence of the four C–H bonds in methane with the outer shell electron configuration, $2s^2 2p^2$, of carbon? Hydrogen contributes its one electron in a 1s orbital, whereas carbon has four atomic orbitals of appropriate energy available for bonding. These four orbitals are one 2s and three 2p orbitals, which are designated $2p_x$, $2p_y$ and $2p_z$ according to their different directions in space; x, y and z refer to the directions of a Cartesian coordinate system. A 2s and a $2p_y$ orbital are shown in Figure 1.1.

The 2s orbital is spherically symmetrical, and its electron density is highest at the nucleus. Each 2p orbital is cylindrically symmetrical

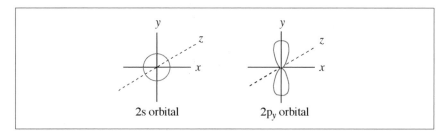

Figure 1.1

and, in contrast to its 2s counterpart, has zero electron density at the nucleus. If one 2s and three 2p orbitals were used without modification in bonding, then four equivalent bonds would not be found in methane.

A slightly higher energy state of carbon is described by the electron configuration $1s^2 2s 2p^3$, and from the 2s and the three 2p orbitals four equivalent mixed or 'hybrid' orbitals are obtained. In the case of carbon in methane, and saturated carbon in most other molecules, the hybridization is described as sp^3. These orbitals have some characteristics of the component atomic orbitals. Accordingly, an sp^3 orbital has finite charge density at the nucleus, as does the 2s (but not the 2p) orbital; also it has directionality, in common with the 2p orbitals, with lobes now of unequal size.

The structure of methane that indicates the equivalent sp^3 hybrid orbitals of carbon bonding with four hydrogens is shown in Figure 1.2. The observed HCH bond angles in methane are such as to place the hydrogen atoms as far apart as possible. The same angles are observed in, say, CF_4, but in general the C–C–C bond angles in acyclic saturated molecules are slightly greater, and an example is propane where the value is *ca.* 112°.

Four equivalent C–H bonds are formed from sp^3 hybrid orbitals, themselves derived from one 2s and three 2p orbitals.

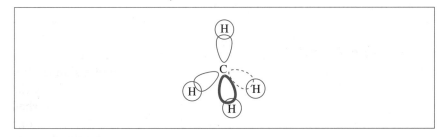

Figure 1.2

1.3 Hybridization: Ethene and Alkenes

Analysis of the geometry of methane does not explain that of ethene, C_2H_4, which contains two equivalent carbons. From electron diffraction data (see Thompson[1]), ethene is known to be a flat molecule with all six atoms in a plane and with bond angles of 120°. The bonding of ethene can be rationalized by using the same orbitals, one 2s and three 2p ($2p_x$, $2p_y$, $2p_z$), but with the difference that *one* of the 2p orbitals does not participate in the hybridization. The net result is that there are three sp^2 hybrid orbitals and one p orbital per carbon. There are a number of significant differences about bonding in ethene compared to that in methane. The sp^2 hybrid orbitals in ethene and indeed all alkenes possess a greater percentage 's' character, and consequently the electrons reside closer to the carbon nucleus. Additionally, the change of hybridization brings about a change of bond angle. In ethene, and around the sp^2

Three sp² bonding orbitals are formed from one 2s and two 2p orbitals. The remaining p orbital on each carbon of ethene is used to form the π bond.

hybrid carbons of alkenes, the bond angles are 120°, which leads to the flat structure. This geometrical arrangement means that the substituents around the alkene carbons are as far away from each other as possible, a feature that is true also of methane (see Section 1.2). It follows that in cyclohexene, for instance, four of the six carbon atoms (and two hydrogens) are coplanar.

What then of the p orbitals that did not participate in the formation of sp² hybrid orbitals? These remain as p orbitals, one at each carbon, and each p orbital contains one electron. Together they form a bond, termed a π bond, between the carbon atoms. The mode of overlap is 'sideways on'. This contrasts with the end-on overlap that results in formation of σ bonds, and which in the present context involves C–H bonds in methane (Figure 1.2) and ethene, and the other bond between carbon and carbon in ethene.

The sp² hybrid orbitals in ethene and the p orbital are shown in Figure 1.3. The sp² hybrid orbitals form one carbon–carbon σ bond and two carbon–hydrogen σ bonds. Ethene may be drawn as in Figure 1.4a; in Figure 1.4b ethene is drawn to highlight the overlap of the p orbitals of the two carbons that give rise to the π bond.

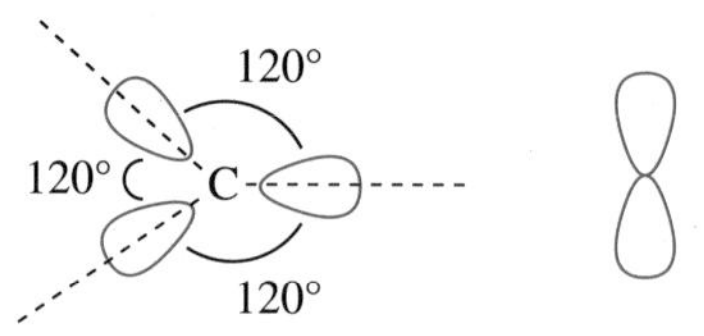

Figure 1.3

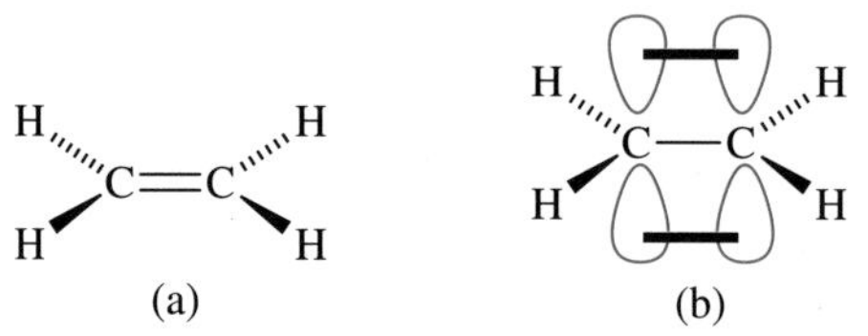

Figure 1.4

The presence of the π bond confers properties on an alkene that mark it out as different from an alkane. In particular, the π bond, by the nature of its sideways overlap of the constituent p orbitals, is weaker than a σ bond. Moreover, the electrons of the π bond are relatively exposed, above and below the plane of the alkene. These electrons are the source of reactivity of the alkene toward electrophiles, as in, say, electrophilic addition of bromine (Chapter 4). The π bond in ethene (and other alkenes) is, however, sufficiently strong that it prevents rotation around the carbon–carbon σ bond, which is a well-documented property of the carbon–carbon bond in ethane (Section 1.6). The bonding between sp²

The π bond of alkenes reacts with electrophiles, *e.g.* bromine.

hybrid carbons of an alkene consists of one σ bond and one π bond; collectively, these are referred to as a double bond.

Double bonds are also formed between carbon and nitrogen to give imines ($R_2C=NR$), and also between carbon and oxygen to give aldehydes and ketones. In both these cases also there is one σ bond and one π bond. In aldehydes, ketones and imines, oxygen especially, and nitrogen, are more electronegative than carbon. Accordingly, the electrons of the π bond in particular are drawn toward the oxygen and the nitrogen, respectively. This has the effect of making oxygen and nitrogen partially negatively charged, and the carbons of these double bonds are correspondingly partially positive.

1.4 Hybridization: Ethyne

For completeness, we consider briefly the triple bond in ethyne (acetylene, HC≡CH), though its stereochemical significance is limited. The carbons of ethyne utilize one s and one p orbital to form two sp hybrid orbitals. This leaves two unhybrid p orbitals. The sp hybrid orbitals are used to form the σ bonds, one to the other carbon, and one to hydrogen. The p orbitals at each carbon form two separate, and independent, π bonds that are at right angles to each other as viewed along the axis of the molecule. Such π bonds are called orthogonal. Bonding in nitriles, RC≡N, is similar with the triple bond again made up of one σ bond and two orthogonal π bonds. The bonding in ethyne is shown diagrammatically in Figure 1.5.

Ethyne sp hybrid orbitals form two σ bonds; two remaining p orbitals on each carbon form orthogonal π bonds.

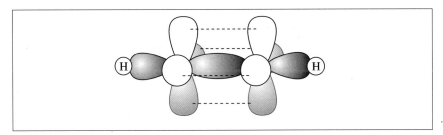

Figure 1.5

Along the series ethane (carbons sp^3), ethene (sp^2) and ethyne (sp) there is an increase in the percentage s character of the carbon hybrid orbitals. This implies a greater electron density at the carbon nucleus as one moves along this series, which is in accord with the progressively shorter carbon–carbon bond lengths observed. Average values of 0.154 nm (ethane), 0.133 nm (ethene) and 0.12 nm (ethyne) have been found. Table 1.1 lists some bond angles in hydrocarbons.

Table 1.1 Bond angles in hydrocarbons[a]

Bond type	Bond angle
C–C–C	109°28' (ideal)
	111–112° (found)
C=C–H	120°
C≡C–H	180°

[a]Values can vary by a few degrees from those cited

1.5   Bonding and Anti-bonding Orbitals

A further general point about molecular orbitals is illustrated with the hydrogen molecule. This is composed of two hydrogen atoms, each of which contributes one electron from a 1s orbital to form the molecular orbital. Two electrons in this molecular orbital form the σ bond in the hydrogen molecule. It may appear that two atomic orbitals, that together are *capable* of containing four electrons, form one molecular orbital that is only capable of containing two electrons. The combination of atomic orbitals is slightly more involved.

When two atomic orbitals combine they produce two molecular orbitals of unequal energy (Scheme 1.1). The lower energy molecular orbital is lower in energy than the component 1s atomic orbitals; it is called a bonding molecular orbital (MO) and contains the two electrons that form the σ bond. Additionally, a second molecular orbital is formed that is higher in energy than the 1s orbitals, and in the ground state hydrogen molecule it is unoccupied. This higher energy orbital is known as an anti-bonding orbital and in this case is given the symbol σ*.

When two atomic orbitals produce two molecular orbitals, one is lower in energy (bonding), and the other is higher in energy (anti-bonding) than the atomic orbitals.

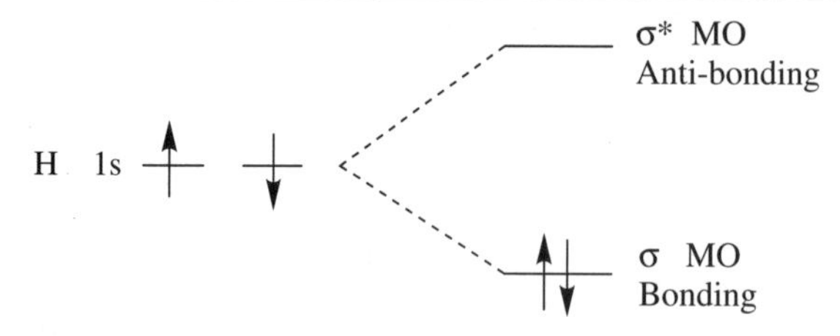

Scheme 1.1

Similar behaviour is also shown by p orbitals, which form π and π* molecular orbitals.

1.6   Conformation: Ethane

Unlike methane with its fixed geometry, ethane is not rigid and rotation occurs around the carbon–carbon bond. Conformations of a molecule are three-dimensional arrangements that differ *only* by rotation around a single bond. Each increment of rotation, however small, produces a change of conformation. In particular, the atoms remain connected in the same order during conformational change, with bonds being neither made nor broken. To explore conformation it is necessary to define terms and to examine ways of representing molecules, or their key parts, in two dimensions. Molecules containing saturated carbon will be considered here.

Ethane can have a continuous series of conformations as rotation proceeds around the carbon–carbon bond. In order to analyse this rotation it is appropriate here to present two different conventions for representing ethane, and other molecules, so that their three-dimensional character can be appreciated. One is the Newman projection **2**, in which ethane is viewed along the carbon–carbon bond. The C–H bonds of the 'front' methyl group are joined to the centre of the circle, which signifies the carbon–carbon bond axis. The corresponding C–H bonds of the rear CH_3 group are differentiated in that they finish at the circumference of the circle. In the eclipsed conformation the dihedral angle H^a–C(1)–C(2)–H^b is 0° (carbon atom numbering as in diagram **3**). The dihedral angle is the angle between planes H^a–C(1)–C(2) and C(1)–C(2)–H^b, but is conveniently referred to as above. The dihedral angle is sometimes simply known as the torsion angle.

Conformational changes are made by rotation around bonds. No bonds are broken or made.

The second representation is known as a sawhorse and is shown in **3**. Here the molecule is observed from an oblique angle.

Small three-dimensional molecules are commonly viewed by sawhorse or Newman projections.

> Make a molecular model of ethane and confirm that **3** represents an eclipsed conformation.

As the C–C bond in ethane rotates about its axis the value of the dihedral angle ϕ increases gradually. When ϕ is 60°, the conformation is termed staggered, and the hydrogens are now as far apart as possible. Accordingly, this conformation corresponds to an energy minimum and is represented by **4** (Newman) and **5** (sawhorse) projections. Hydrogens H^a and H^b (and also the other pairs) are as close as possible in the eclipsed conformation, which therefore represents an energy maximum.

A clockwise rotation of 120° of the rear methyl group from the eclipsed position in **2** and **3** gives a second eclipsed conformation identical with

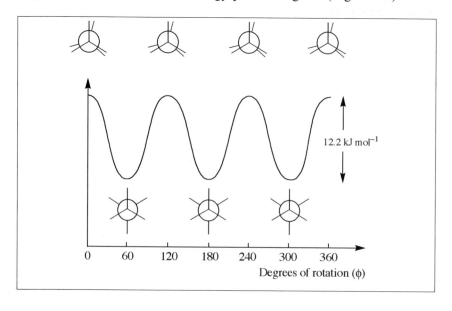

4

5

6

7

8

2 and **3**, and therefore with the same energy. A further 120° clockwise rotation gives a third identical conformation. These are shown in saw-horse projection in **6** and **7**, respectively. A still further 120° clockwise rotation completes the revolution, with the original pair of eclipsed hydrogens H^a and H^b now again coincident.

The conformation shown in **8** represents one of a number of skewed conformations in which the rotation around the C–C axis, from the position shown in **2** (or **3**), has occurred such that the hydrogens of the rear methyl group have moved in a clockwise direction by an amount $0 < \phi < 60°$. This rotation around the carbon–carbon bond in ethane is rapid at room temperature, and is sometimes described as 'free rotation'. This is not strictly true as a small, though definite, energy barrier is encountered.

We now consider the energy profile of an ethane molecule as the carbon–carbon bond is tracked through one revolution. For our purposes the front CH_3 group in the Newman projection is kept constant and the rear CH_3 group rotates as the C–C bond itself rotates around its axis. As this happens the H^a–C(1)–C(2)–H^b dihedral angle varies, and this angle forms the x-axis in the energy profile diagram (Figure 1.6).

The barrier to rotation around the C–C bond in ethane has three-fold periodicity. Three conformations, in order of decreasing energy, are eclipsed, skewed and staggered.

Figure 1.6 Torsion strain energy versus bond rotation in ethane

It is clear from Figure 1.6 that eclipsed conformations are of highest energy, skewed conformations are of intermediate energy, and staggered conformations are the most stable. The energy difference between the highest and lowest energy conformations is 12.2 kJ mol^{-1} at 25 °C, and is referred to as the **torsion barrier**. In more complicated molecules than ethane, *e.g.* C_{abc}–C_{xyz}, the rotational barrier will differ from that shown in Figure 1.6; the peaks and troughs are now all of unequal height and depth and, depending on the substituents, the energy profile may only be repeated once per revolution.

1.7 Conformation of Propane and *n*-Butane

Consideration of the Newman projection of propane (**9**) suggests that the energy profile associated with rotation around the CH_3–CH_2 bond contains both three identical maxima and three identical minima per revolution. In eclipsed conformations there is now a non-bonded repulsion between a methyl and a hydrogen; accordingly, the torsional barrier, at 14.2 kJ mol^{-1}, is a little higher than in ethane.

The rotational barrier in propane has three-fold periodicity.

The case of butane, $C(1)H_3$–$C(2)H_2$–$C(3)H_2$–$C(4)H_3$, is more involved. Three conformations are shown in **10–12**, in which the methyl groups are, for convenience, represented by X. In **10** the $C(1)$–$C(2)$–$C(3)$–$C(4)$ dihedral angle, ϕ, is 0° and this corresponds to the *cis* conformation of butane. If one rotates the rear group clockwise, then when $\phi = 60°$ a staggered conformation **11** is obtained; this is known as a *gauche* conformation. A further 120° clockwise rotation of the rear group leads, *via* an eclipsed conformation, to **12**, known as *anti*, since the dihedral angle, ϕ, is now 180°. In butane this conformation has a plane of symmetry. A further 120° clockwise rotation yields a *gauche* conformation **13**, that is equal in energy to **11**.

Butane has two *gauche* conformations in which the dihedral angle between the CH_3 groups is 60°, and an *anti* conformation, which is the most stable, with a corresponding dihedral angle of 180°.

cis
10

gauche
11

anti
12

gauche
13

In the vast majority of appropriately constructed molecules, for example derivatives of butane, the *anti* conformation, *e.g.* **12**, in which the groups X are each flanked by two hydrogens, is the most stable; however, there are a few exceptions. The *anti* conformation of butane is more stable than the *cis* by *ca.* 18.8 kJ mol^{-1}. An energy profile of rotation around the central C–C bond of butane is given in Figure 1.7.

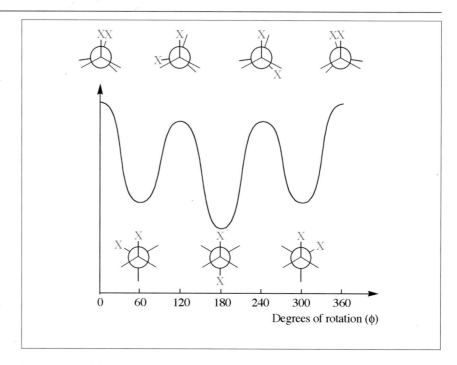

Figure 1.7 Torsion strain energy versus bond rotation about the central, C(2)–C(3), bond in butane

> With the aid of models, verify that the *cis*, *gauche* and *anti* conformations have the relative energies indicated by the above data.

The conformations of longer, unbranched hydrocarbons become more complicated the greater the number of carbons. For example, pentane has nine staggered conformations.

1.8   Cyclohexane: Chair Conformation

Cyclohexane is a saturated cyclic hydrocarbon, C_6H_{12}, in which all the carbons are sp^3 hybridized and tetrahedral.

> Use models to show that cyclohexane cannot adopt a planar structure.

Benzene, C_6H_6, is an unsaturated hydrocarbon with a six-membered ring in which all carbon atoms are now sp^2 hybridized.

> Make a model to show that benzene is flat.

The cyclohexane ring, either alone or as part of a more complex structural unit, occurs in certain natural products and, accordingly, cyclohexane is the most important saturated cyclic hydrocarbon. Two principal conformational isomers exist. The more stable is called the **chair conformation 14**, and the less stable the **boat conformation** (see Section 1.9). In both these conformations the C–C–C bond angles are close to the tetrahedral value of 109°28′; consequently, cyclohexane has little **angle strain**. Angle strain becomes significant in saturated hydrocarbons if there are meaningful departures from the above value.

The chair conformation of cyclohexane also has minimum torsion strain, as can be confirmed both from a Newman projection and molecular models. The Newman projection **16**, derived from **15**, in which one looks along the C(1)–C(2) bond of cyclohexane, reveals an almost exactly staggered local conformation. This situation is reproduced if one inspects the Newman projections along the other five carbon–carbon bonds in chair cyclohexane. Inspection of **14** indicates that of the 12 hydrogens in a molecule of cyclohexane, six are parallel to a three-fold axis of symmetry (C_3 axis) that passes through the molecule as indicated. These six hydrogens are therefore termed **axial**, and are further differentiated in that three are above an approximate plane through the molecule and three are below. A molecular model of cyclohexane will accordingly rest on a table by either the three upper or the three lower hydrogens.

Chair cyclohexane has low angle and torsion strain, and a three-fold axis of symmetry. With respect to this axis, six hydrogens, termed 'axial', are parallel, and six hydrogens, termed 'equatorial', are approximately perpendicular.

1,3-Diaxial hydrogens are *cis*; 1,2-diaxial hydrogens are *trans*.

The three upper axial hydrogens in cyclohexane are termed *cis* to each other; likewise the three lower axial hydrogens are mutually *cis*. These *cis* relationships are at the origin of the phrase '*cis* 1,3-diaxial interaction' that is used in appropriate 1,3-disubstituted cyclohexanes (Chapter 6). It may be helpful to insert an imaginary C(1)–C(3) bond to confirm with the aid of a model that the dihedral angle is 0°.

It is noteworthy also that axial hydrogens on adjacent carbons, *e.g.* H^{1a} and H^{2a} in **16** and **17**, are said to be *trans* by virtue of the dihedral angle H^{1a}–C(1)–C(2)–H^{2a} being 180°. These relationships are quite general whether the relevant sites are occupied by hydrogen or another substituent(s). Relationships between hydrogens in other positions of the cyclohexane ring are considered later (Chapter 6).

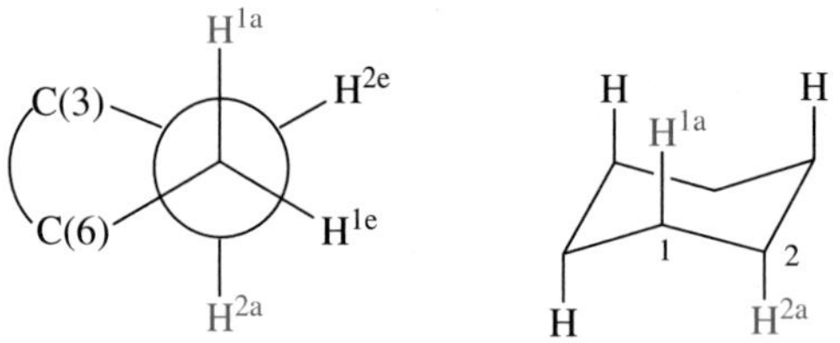

16   17

The remaining six C–H bonds of cyclohexane are more or less perpendicular to its C_3 axis, and are termed **equatorial** bonds. Inspection of a model will confirm that they are directed away from the core of the molecule and do not experience significant non-bonded interactions.

As well as familiarity with a molecular model of cyclohexane, it is important to be able to present a good drawing; this applies both to **14** and to the ring inverted form (see Section 1.9).
(i) In a perspective drawing, a C–C bond is drawn parallel to the corresponding bond across the ring, *e.g.* the C(1)–C(2) and C(4)–C(5) bonds must be parallel.
(ii) If included, the six axial C–H bonds are all drawn mutually parallel.
(iii) Importantly, each equatorial C–H bond forms a dihedral angle of 180° with two ring C–C bonds; when this dihedral angle occurs the bonds are said to be *trans* to two ring C–C bonds. In **15**, therefore, C(2)–H^{2e} is *trans* to the C(1)–C(6) and C(3)–C(4) bonds since the dihedral angles H^{2e}–C(2)–C(1)–C(6) and H^{2e}–C(2)–C(3)–C(4) are both 180°. With the aid of a molecular model check that this is so. The above bonds should therefore always be drawn parallel to each other. Remember that this applies also to other equatorial C–H bonds and the corresponding ring C–C bonds in cyclohexane.

1.9   Cyclohexane: Boat Conformation

Boat cyclohexane contains (a) two sets of high-energy butane-like eclipsed interactions and (b) a cross-ring 1,4 hydrogen–hydrogen interaction, and is thus less stable than chair cyclohexane.

The boat conformation of cyclohexane (**18**) can be constructed from a molecular model of the chair form by holding the right-hand three carbons C(2), C(3) and C(4) of **15**, clamped from the top with the hand and moving the left-hand three carbons upward. A Newman projection of the boat form looking along the C(1)–C(2) bond, and shown in **19**, is reminiscent of the highest energy *cis* conformation of butane.

18   **19**

> With the aid of models confirm:
> (i) that a similar eclipsed *cis* conformation is found around the 'opposite', C(4)–C(5), ring bond
> and
> (ii) that, in contrast, staggered conformations exist around the other four ring C–C bonds.

Further, the boat conformation of cyclohexane is usually thought to possess an unfavourable transannular non-bonded interaction between the two hydrogens marked H′ in **18**; this is known as a '1,4-interaction'. Recently it has been claimed that the magnitude of the 1,4-interaction in **18** is negligible, but that there is a 1,4-repulsive interaction between C(1) and C(4) in the boat conformation (see Sauers[2]).

On account both of this interaction and the two sets of eclipsed interactions the boat conformation is less stable than the chair counterpart by 27.5 kJ mol^{-1}. A further conformation of cyclohexane, the twist-boat, is known and is relevant in ring inversion (Section 1.10), but is not considered further here. It is mentioned briefly in Chapter 6.

An appreciation of the 1,4-interaction is best achieved if space-filling models are used. 'Normal' molecular models, though excellent for an understanding of conformations, often under-represent the 'size' of atoms. For example, in Fieser models of cyclohexane, hydrogen atoms are not included, and hence no account is taken of the size of hydrogen and the 'size' of carbon in, for example, cyclohexane is under-represented, except in space-filling models. The size of an atom is given by its van der Waals' radius. When another atom intrudes into the space of an atom, *i.e.* closer than its van der Waals' radius, a repulsion occurs and this is destabilizing.

The van der Waals' radii of a number of common groups are given in Table 1.2. Space-filling molecular models take account of the van der Waals' radii and should be used when matters of non-bonded interaction are involved or suspected.

Table 1.2 van der Waals' radii of common groups

Group	van der Waals' radius (nm)
H	0.12
CH_2	0.20
CH_3	0.20
F	0.135
O	0.14
N	0.15
Cl	0.18
Br	0.20
I	0.22

1.10 Inversion of Cyclohexane

^{1}H Nuclear magnetic resonance (NMR) spectroscopy is a sensitive technique in which the absorption (at higher or lower fields) of different protons in a molecule reflects the different environment of the protons. However, the ^{1}H NMR spectrum of cyclohexane at ambient temperature shows only a single absorption. The reason for this is that cyclohexane is interconverting rapidly between two chair conformers of equal energy. This is shown by the equilibration of **20** and **21** in which only one pair of geminal hydrogens, H^1 and H^2, is shown. After inversion an axial hydrogen, H^1 in **20**, has become equatorial in **21** and *vice versa*. This behaviour is quite general for all hydrogens in cyclohexane.

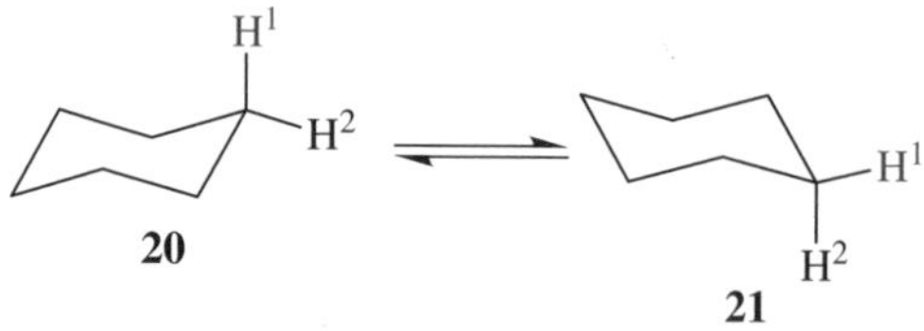

After inversion of chair cyclohexane, axial hydrogens have become equatorial and *vice versa*; this inversion is rapid at room temperature.

At room temperature, inversion of cylcohexane is so rapid that its ^{1}H NMR spectrum shows only a single averaged absorption for the 12 protons. At low temperatures, less than 230 K, when the rate of interconversion of **20** and **21** is slow, it is possible to observe separate absorptions for the axial and equatorial protons.

If one examines the ^{1}H NMR spectrum of cyclohexane at temperatures in the range in which the changeover takes place from a single average absorption to separate axial and equatorial absorptions, it is possible to estimate the rate constant for the ring inversion of cyclohexane. One can translate this result to room temperature and say that here cyclohexane is undergoing ring inversion more than 100,000 times per second. For further details see Eliel and Wilen[3] (and the references cited therein).

Ring inversion of cyclohexane occurs by a route that involves twist-boat, and possibly boat, conformations and which is not reproduced accurately in the manual inversion outlined below.

Practise this ring inversion of chair cyclohexane with the aid of models. Adopt the protocol described in Section 1.8 for formation of a boat conformation of cyclohexane. Then hold carbons C(6), C(5) and C(1) clamped in the hand, and rotate the right-hand three carbons downward to complete the inversion. It is helpful to make two identical models with H[1] and H[2] marked by, for example, different coloured tape or similar. Keep one model for reference, and perform the inversion on the other.

1.11 Monosubstituted Cyclohexanes

The most stable form of a monosubstituted cyclohexane is, like cyclohexane itself, a chair conformation. There are, however, two valid chair conformations and again, like cyclohexane, these are interconvertible by ring inversion. In the particular case of methylcyclohexane, they are shown by **22** and **23**.

It can be seen that in **22** there are two destabilizing *cis* 1,3-diaxial interactions between Me and H. The molecule responds to these interactions by undergoing ring inversion to produce **23** in which the methyl group is now in the relatively 'open' equatorial position. At 25 °C the equilibrium ratio of **23** to **22** is *ca.* 18:1, and this corresponds to a free energy difference of $\Delta G° = -7.1$ kJ mol^{-1}.

The equilibrium constant, K, is related to the standard free energy, $\Delta G°$, by equation (1). In Table 1.3, values of $\Delta G°$ for a number of values of K are presented; it will be useful to become familiar with the range of values of $\Delta G°$ involved. For example, if K is 9, the derived value of $\Delta G°$, *ca.* -5.5 kJ mol^{-1}, is quite modest.

$$\Delta G° = -RT\ln K \qquad\qquad (1)$$

Table 1.3 Relationships between free energy and equilibrium constants

% More stable isomer	K	$-\Delta G^\circ$ (kJ mol^{-1})
50	1	0
55	1.22	0.49
60	1.50	1.01
65	1.86	1.54
70	2.33	2.10
75	3.00	2.72
80	4.00	3.44
85	5.67	4.30
90	9.00	5.44
95	19.00	7.29
99	99.00	11.39
99.9	999.00	17.12

Conformational inversion of monosubstituted cyclohexanes results in equilibration of the substituent between axial and equatorial positions. The bulkier a substituent, the more it prefers the equatorial position. The 1,1-dimethylethyl group occupies the equatorial position exclusively.

In 1-methylethylcyclohexane (isopropylcyclohexane) the conformer with the equatorial substituent (shown in **24**) is now favoured over its axial counterpart by a factor of *ca.* 35 because of the larger size of the alkyl substituent, and this ratio corresponds to $\Delta G^\circ = -8.8$ kJ mol^{-1}.

24   **25**

Two more examples merit consideration. In 1,1-dimethylethylcyclohexane (**25**, *t*-butylcyclohexane) the very bulky substituent group has an overwhelming preference for the equatorial position on account of its severe *cis* 1,3-diaxial interaction with two hydrogens when it is axial; this can be verified with space-filling models. In effect, **25** is locked in the chair conformation with the 1,1-dimethylethyl group equatorial, incapable of inversion; this substituent is discussed further in Chapter 6.

By way of contrast, in iodocyclohexane, iodine has little preference as to whether it is axial or equatorial. The reason is that although iodine is a large atom, the C–I bond length of *ca.* 0.195 nm is much longer than the axial C–H bonds (*ca.* 0.11 nm), and although the van der Waals' radius of iodine, *ca.* 0.22 nm, is large, 1,3-diaxial interactions that involve iodine are not significant.

A list of conformational preferences of substituents for the equatorial position in monosubstituted cyclohexanes, expressed in terms of ΔG°, and often known simply as *A* values, is available (see Bushweller[4]) and a selection of *A* values is presented in Table 6.1.

A note on diagrams is relevant here. Hydrogens are either shown explicitly, usually to demonstrate a point, as in **22** and **23**, or they may be omitted, again as in **22** and **23**. It is therefore in order to show some hydrogens, but not others, in a diagram of a given molecule. An un-annotated line in a structure represents a methyl group. Examples of this are in the 1-methylethyl group in **24** and the 1,1-dimethylethyl group in **25**. Of course, the methyl groups may be written as in **22** and **23**. Diagram **23** may equally well be drawn as in **26**. All other groups, *e.g.* Cl, NO_2, are annotated.

26

Summary of Key Points

- The conventional representation of methane in two dimensions is given, and is applicable to saturated carbon in general.
- In methane, four equivalent C–H bonds arise from hybridization of one 2s and three 2p carbon orbitals to give four equivalent sp³ hybrid orbitals.
- The HCH bond angles at carbon in methane are 109°28′ (in most saturated hydrocarbons, values of *ca.* 111° are found for CCC bond angles).
- Both carbons in ethene are sp² hybridized. The component orbitals that form the hybrid orbitals are one 2s and two 2p. There remains a p orbital on each carbon, and these orbitals (each of course containing one electron) overlap sideways to form a π bond.
- Rotation around the axis of the C–C σ bond in ethane means that one set of three hydrogens rotates relative to the other. When this happens the conformation of ethane changes.
- As hydrogens pass each other (eclipsed conformation) the energy of the ethane molecule rises.
- At 60° away from this eclipsed conformation comes the most stable conformation, called staggered; intermediate conformations are called skewed.
- As rotation proceeds the energy of ethane therefore varies in a pattern that repeats itself three times per revolution. The energy profile associated with rotation about a C–C bond in propane likewise shows a three-fold periodicity.
- The more involved rotation around the central C–C bond in butane contains a conformation called *gauche*, in which the dihedral angle between the methyl groups is 60°. When this angle is 180° the conformation is *trans*.
- Cyclohexane, the most important cyclic hydrocarbon, is most stable in the chair conformation, which has very little angle and torsion strain. Hydrogens occupy equatorial and axial positions.

- At room temperature, cyclohexane inverts conformation rapidly, and in so doing the equatorial hydrogens become axial and correspondingly the axial hydrogens become equatorial.
- 1,3-Diaxial hydrogens are mutually *cis*, and 1,2-diaxial hydrogens are mutually *trans*.
- The boat conformation of cyclohexane is less stable because the local conformations around two of the six C–C bonds are eclipsed; there is a non-bonded repulsion between carbons C(1) and C(4), in addition to a possible 1,4-transannular non-bonded repulsive interaction between two hydrogens.
- In monosubstituted cyclohexanes the chair conformation is again more stable, and the substituent can be either equatorial or axial.
- Substituents prefer the more open equatorial position, free from non-bonded interactions, to an extent that is greater the bulkier the substituent.

References

1. H. W. Thompson, *Trans. Faraday Soc.*, 1939, **35**, 701.
2. R. R. Sauers, *J. Chem. Educ.*, 2000, **77**, 332.
3. E. L. Eliel and S. H. Wilen, *Stereochemistry of Organic Compounds*, Wiley, New York, 1994, p. 597.
4. C. H. Bushweller, in *Conformational Behaviour of Six-Membered Rings*, ed. E. Juaristi, VCH, Weinheim, Germany, 1995, p. 25, and references cited therein.

Further Reading

D. H. R. Barton, *The Principles of Conformational Analysis* in *Nobel Lectures: Chemistry 1963–70*, Elsevier, Amsterdam, 1972, p. 298.

J. March, *Advanced Organic Chemistry*, 4th edn., Wiley-Interscience, New York, 1992, pp. 3 and 146.

J. McMurry, *Organic Chemistry*, 4th edn., Brooks-Cole, Belmont, California, 1996, p. 106.

2
Chiral Molecules: One Stereogenic Centre

Aims

By the end of this chapter you should be familiar with:

- All the terms in coloured bold type
- The relationship between enantiomers
- The principles of polarimetry
- Fischer projections
- The CIP *R/S* configuration nomenclature
- Racemization

2.1   Chirality, Enantiomers and Optical Activity

When an sp^3 hybridized tetrahedral carbon is attached to four different groups, as in **1** (see below), the molecule cannot be superimposed on its mirror image. This is the same property that a right hand, say, placed in front of a plane mirror possesses. Molecules such as **1** are said to be chiral (or handed), from the Greek word 'cheir' for hand.

Chirality then is the property of handedness, and the adjective 'chiral' refers to the molecule as a whole, rather than to a particular atom. Molecules such as CCl_4 and CH_2Cl_2 that can be superimposed on their mirror images are said to be 'achiral'.

Molecules such as **1** and **2** are termed 'enantiomers', from the Greek 'enantio' meaning opposite. Enantiomers are defined as a pair of molecules related as *non-superimposable* mirror images; note that it is essential to include 'non-superimposable' in the definition. It can be seen that enantiomers are a particular sub-class of stereoisomers. Each enantiomer of a pair has the same physical properties (*e.g.* m.p., b.p. and solubility) with one exception. This distinguishing feature arises when, in separate experiments, plane polarized light is passed through a solution of each enantiomer in the same solvent, and using the same cell. If the concentration of the two enantiomers is the same, then the plane of polarized light is rotated in opposite senses and by the same amount. This difference in behaviour has important diagnostic value.

Although this chapter deals with chirality in certain molecules, one should be aware that chirality is to be found in the wider world. As mentioned above, the word chirality comes from the Greek for hand. A hand, and a foot, are chiral and fulfil the same criteria as chiral molecules. Accordingly, a left hand placed in front of a mirror gives a right hand as its mirror image (Figure 2.1). These cannot be superimposed, as can readily be demonstrated with one's own hands. Also, a glove is chiral and a pair of gloves consists of two enantiomeric gloves. Likewise, a shoe or a sandal is chiral. A plain sock is not chiral but it will of course readily adapt to the chirality of the foot.

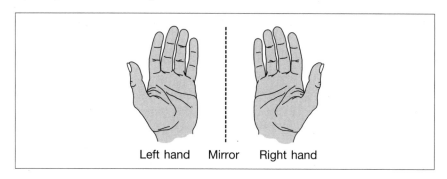

Left hand   Mirror   Right hand

Figure 2.1

Another common chiral object is a cylindrical helix, such as a bedspring. A helix with its mirror image is shown in Figure 2.2 and again object and mirror image are not superimposable. The thread of a screw ensures that it too is chiral. The body shell of a car as it enters the assembly line is not chiral, and it has a plane of symmetry. When the car leaves the assembly line it has acquired, for example, a steering wheel on one side or another. The finished car is now not superimposable on

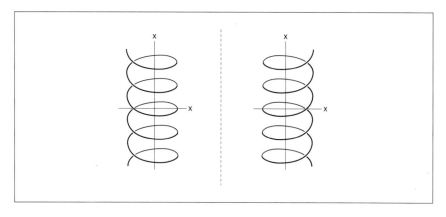

Figure 2.2

its mirror image, and it is therefore chiral; it has also lost its plane of symmetry.

A rowing boat is clearly not chiral, but note that it possesses a plane of symmetry. An engine, presumed symmetrical, fitted to this boat will have a propeller, almost certainly with three blades. The propeller itself is chiral (verify this by observation if possible) and after the engine with its propeller is fitted, the boat as a whole has now become chiral by virtue of its propeller, which has also caused the boat to lose its plane of symmetry.

Some types of aircraft use one or more propellers that each have four blades; each propeller is again chiral. Is a jet engine chiral? (Consider the engine, not running, viewed from the front of the plane).

Light may be regarded as a wave motion that contains oscillating electric and magnetic fields. The electric field of light oscillates in all planes at right angles to the direction of propagation of the light wave. When a beam of normal light is passed through a device called a **polarizer**, that in effect acts as a filter, only light waves that are oscillating in a single plane are allowed to pass through. The polarizer thus serves to block the passage of light that is oscillating in all other planes. The light that emerges from the polarizer is said to be plane polarized.

In 1815 the French physicist Biot made the important observation that when a beam of **plane polarized light** is passed through a solution of certain naturally occurring organic compounds, the plane of polarized light is rotated either to the left or to the right. Molecules that induce this rotation are said to be **optically active**.

The simplest class of optically active molecules has one carbon bonded to four different substituents, and of course it is necessary for one enantiomer to be in excess over the other. Measurements of optical activity are carried out with a polarimeter, which is shown schematically in Figure 2.3. The direction and extent of the rotation of the plane of polarized light is the basis of specific rotation, which is dealt with in this section.

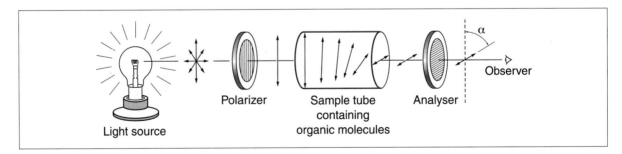

Each enantiomer undergoes chemical reactions at the same rate as long as the other reagent (and the solvent) is not chiral. However, in the

Figure 2.3

Enantiomers are chiral molecules that are non-superimposable mirror images of each other, and with one exception have identical physical properties. Enantiomers react at the same rate with achiral reagents.

case in which the other reagent is chiral, the two enantiomers will react at different rates; this is the basis for enzymic selectivity, which is discussed later (Section 3.8).

A pair of enantiomers is represented by **1** and **2**, in which the carbons shown in colour are called 'stereogenic centres'. They are also less justifiably called 'chiral centres'; the term 'asymmetric carbon' is now encountered less frequently.

The opposite configurations of the stereogenic centres in **1** and **2** are indicative of the opposite directions in which they rotate the plane of polarized light. However, it is not possible from inspection of a particular enantiomer to say in which direction the plane is rotated. Enantiomers that rotate the plane to the right, *i.e.* clockwise, are called **dextrorotatory** and are sometimes given the symbol '*d*'; those that rotate to the left, anti-clockwise, are correspondingly called **laevorotatory** with the symbol '*l*'. It should be made clear that the terms dextrorotatory and laevorotatory are defined with the observer looking into the propagating beam. These names and their symbols are empirical and have no fundamental significance. Also, those compounds that rotate the plane of polarized light to the right are frequently given the symbol (+), and those that rotate to the left the symbol (−).

In order to characterize the extent of rotation, an index called specific rotation, $[\alpha]_D$, is used; this is given by equation (1):

$$[\alpha]_D^t = \frac{100\alpha}{l \times c} \tag{1}$$

In equation (1), 'α' (alpha) refers to the observed rotation in degrees, '*l*' to the length of the polarimeter cell in decimetres, and '*c*' refers to the concentration of the sample in g 100 cm^{-3}, *i.e.* g per 100 cm^3. On the left-hand side of the equation, '*t*' refers to the temperature in °C at which the measurement takes place, and the subscript 'D' indicates that light from the D line of a sodium lamp with a wavelength of 589.6 nm was used. Sometimes the concentration c is expressed as g cm^{-3}; in this case, equation (2) is employed:

$$[\alpha]_D^t = \frac{\alpha}{l \times c} \tag{2}$$

If a measurement is made as a check on a published value of the specific rotation, care must be taken to ascertain whether the concentration relates to equation (1) or equation (2). For example, a reasonable concentration, such as 2 g in 100 cm³ in respect of equation (1), might well exceed the solubility of the compound if taken to be concentration $c = 2$ g in 1 cm³, as used in equation (2).

Values of specific rotation are expressed as in equation (3) for the amino acid cysteine (compound **30** in the Problems section of this chapter):

$$[\alpha]_{\mathrm{D}}^{25} = -16.5 \ (\text{conc. 2 g in 100 cm}^3 \ H_2O)$$
$$\text{cysteine} \tag{3}$$

The units of $[\alpha]$ are 10^{-1} deg cm² g⁻¹ but are not usually included; until recently, values were expressed in deg. The conditions used are given because $[\alpha]$ can vary with (a) wavelength (sometimes the incident light is from a mercury lamp and has wavelength 546.1 nm.), (b) temperature and (c) solvent; effects of polarity of the solvent and hydrogen bonding between solvent and substrate can affect $[\alpha]$ markedly.

The use of g cm⁻³ to express concentration in equation (1), though common in polarimetry, is unusual; concentration is more commonly expressed in mol dm⁻³. A molar rotation [M] is known, but is used less frequently.

The example shown by **1**, *i.e.* a stereogenic carbon bonded to four different groups, is the most common reason for a molecule to show chirality, and of course there can be several stereogenic centres in a molecule. However, the need for a stereogenic centre is not an absolute requirement for a molecule to be chiral, and this is considered in Chapter 5. All cases must, however, fulfil the fundamental condition: for a molecule to be chiral and to display optical activity, it must be non-superimposable on its mirror image.

The stereochemistry that is possible at saturated carbon is also discussed by McMurry.[1]

Enantiomers rotate the plane of polarized light in opposite directions by an amount that is characteristic of the compound. Specific rotation, $[\alpha]$, is a measure of the rotation at unit concentration using a cell (that contains the solution) of unit length.

2.2 How to Specify a Configuration

2.2.1 Fischer Projections

Two conventions are in use to describe the configuration of a stereogenic centre. The older system uses the symbols D and L, and is still employed for two important classes of compound, namely amino acids and sugars; otherwise it has been superseded. The D/L system will be considered first. The use of capital letters here serves to distinguish this convention from *d/l* (page 22), which relates only to the sense in which the plane of polarized light is rotated.

In the early 1900s it was not possible to assign an absolute configuration to a chiral molecule with even a single stereogenic centre, although the need for a system of assignment was appreciated. Accordingly, an arbitrary standard was put in place; this consisted of the molecule (+)-glyceraldehyde (**3**), selected because of its close relationship to sugars. The representation of the three-dimensional structure was addressed as follows.

$$CHO \qquad\qquad CHO \qquad\qquad CHO$$

H———OH   H———OH   HO———H

$$CH_2OH \qquad\qquad CH_2OH \qquad\qquad CH_2OH$$

(+)-**3**   **4**   (−)-**5**

(+)-Glyceraldehyde was drawn as **3**, using what is known as a Fischer projection. The convention is that a stereo drawing of this molecule is shown as in **4**. The vertical bonds are directed behind the page from the stereogenic carbon (which is not shown in a Fischer projection) and the horizontal bonds in **3** resemble a bow tie in **4** projecting out of the page. Fischer was the second Nobel laureate in chemistry in 1902.

The decision was taken to assign (+)-glyceraldehyde (**3**) the absolute configuration shown, which was termed 'D', and the enantiomer (−)-glyceraldehyde was then represented by **5**, and given the symbol 'L'. The terms D and L refer to the molecule as a whole. In 1951, Bijvoet and his group in Utrecht showed by X-ray structure analysis that the original arbitrary assignments, happily, were correct. This finding was significant because the configurations of many compounds had been designated by relating them to (+)-glyceraldehyde. One example involves the configuration of (−)-glyceric acid (**6**), obtained by oxidation of (+)-glyceraldehyde [typically with mercury(II) oxide]. Since the oxidation has taken place away from the stereogenic centre and the oxidizing agent is mild, one can conclude that there has been no change in configuration of the stereogenic centre during formation of **6** from **3**.

$$CHO \qquad\qquad\qquad CO_2H$$

H———OH $\longrightarrow$ H———OH

$$CH_2OH \qquad\qquad\qquad CH_2OH$$

(+)-D-**3**   (−)-D-**6**

The configuration of (−)-**6** is as shown and is accordingly termed D-glyceric acid. This result shows that there is no correlation between configuration and the sign of rotation.

Except for glycine, the alpha amino acids derived from proteins of higher living organisms have at least one stereogenic centre. Those with a single such centre are represented by Fischer projection formulae **7** (or the corresponding stereo diagram **8**), and all have the L configuration shown at the alpha carbon. In these amino acids, *e.g.* serine (**7**, R = CH$_2$OH), assignment of configuration was achieved by chemical conversions that related it to L-(−)-glyceraldehyde using L-lactic acid (**9**) as a relay.

Amino acids of type **7** of course may be assigned configurations in the *R/S* convention (Section 2.2.2.); after having read this section, attempt question **2.6** in the Problem section (page 35) to ascertain whether the four amino acids **28–31**, which are all of L configuration, are all described by *R* or *S*.

Fischer projections, used for naturally occurring amino acids and sugars, are associated with symbols D and L. The former indicates the same configuration as the standard, D(+)-glyceraldehyde (**3**).

CO$_2$H   CO$_2$H   CO$_2$H

H$_2$N—┼—H   R—C$^{\prime\prime\prime}$H   HO—┼—H

R   NH$_2$ Me

7   **8**   **9**

There are further rules for the use of Fischer projections: (a) where relevant the vertical direction is that of the carbon chain; (b) in the vast majority of cases the carbon in the highest oxidation state is located at the top of the diagram.

Care must be exercised in testing whether two structures are, or are not, superimposable. Thus, a Fischer projection diagram of a chiral molecule with one stereogenic centre may be rotated only through 180° around the central carbon with preservation of configuration. Rotation of the Fischer projection by 90° inverts the stereogenic centre to give the other enantiomer; exchange of any two groups also produces the other enantiomer. A discussion of the D/L convention is given by Buxton and Roberts.[2]

Verify these rules with a molecular model of **3**.
Verify that if one takes a Fischer projection of, *e.g.* **3**, and exchanges any two groups, one produces the enantiomer.

A limitation of the D/L system may be seen from the example of (+)-camphor (**10**), which was assigned configuration D. Comparison with the formula **3** for D-glyceraldehyde reveals that the inter-relationship is, at best, tenuous. Some suppliers still offer D-camphor for sale.

10

2.2.2 The Cahn–Ingold–Prelog *R/S* convention

Cahn, Ingold and Prelog developed a new convention for assignment of configuration at a stereogenic centre (and other more involved cases). Cahn was editor of the then *Journal of the Chemical Society*, Ingold from University College London initiated and established the study of organic reaction mechanisms, and Prelog from ETH Zurich was a distinguished organic chemist and Nobel laureate in 1975. Frequently referred to by the initials of its proponents, the **CIP convention** (also known as the *R/S* convention) was proposed in 1966,[3] following an earlier paper by Cahn and Ingold in 1951.[4] The convention is now in almost universal use and forms part of the IUPAC (International Union of Pure and Applied Chemistry) rules of nomenclature.[5] In a clean break with previous symbolism the symbols '*R*' (Latin, 'rectus' = right) and '*S*' (Latin, 'sinister' = left) were adopted. Interestingly, the symbol '*R*' was taken from the Latin word for 'right' in the sense of 'just' or 'correct', since the Latin for right (as in direction) is 'dexter', and use of this word would have maintained an undesired continuity with previous nomenclature.

The rules firstly require that the stereogenic centre be specified; next the groups attached to this centre are arranged in a sequence and given numbers 1, 2, 3, 4, corresponding to decreasing atomic number of the atom bonded to the stereogenic carbon. If, in a chiral molecule, two atoms directly bonded to the stereogenic centre are the same, *e.g.* $-CH_2-CH_3$ and $-CH_2-CH_2Cl$, the priority sequence is determined by the relative atomic numbers at the first further point along the chains at which a distinction can be made. The convention thus operates in much the same manner as when one looks up a name in a telephone directory. One can, of course, have a chiral molecule in which the priority rankings are the same at all four carbons bonded to the stereogenic carbon, provided that each substituent is distinct by virtue of its characteristics further away from the stereogenic centre.

In the case of a simple molecule, *e.g.* butan-2-ol (**11**), the relevant priorities are assigned as shown. The molecule is then viewed looking toward the atom of lowest priority (H in **11**) along a projection of the H–C bond, through a 'triangle' constructed of atoms 1, 2 and 3 The sense of decreasing priority $1 \rightarrow 2 \rightarrow 3$ is clockwise for **11**, which has *R* configuration. Strictly speaking, the symbol *R* applies to one carbon but in molecules with one stereogenic centre this is often not specified.

Assignments of configuration are easier if the atom of lowest priority, 4, typically H, is shown behind the plane of the paper. In cases in which the atom with priority 4 is shown in the plane of, or in front of, the paper, extra care is required, and it is beneficial to use a model and re-position the molecule so that it can be re-drawn with the atom of priority 4 behind the plane of the paper.

In the *R/S* convention, substituents around a stereogenic centre are given priority numbers (1, 2, 3, 4) in order of decreasing atomic number. Substituent 1 ranks higher than substituent 2, *etc*.

View the lowest priority substituent, number 4, looking through the stereogenic carbon. If the substituent priority sequence $1 \rightarrow 2 \rightarrow 3$ is clockwise, then the configuration is *R*. If anti-clockwise, the configuration is *S*.

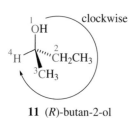

11 (*R*)-butan-2-ol

In a wider context, multiple bonds, such as those in C=O, C=CH$_2$, C$_6$H$_5$, C=NH and C≡N are treated as multiple single bonds and the σ and π bonds are given equal status. These five functional groups become, respectively, as shown in Figure 2.4. Rings are treated similarly; for details see Eliel and Wilen.[6]

Multiple bonds are treated as multiple single bonds.

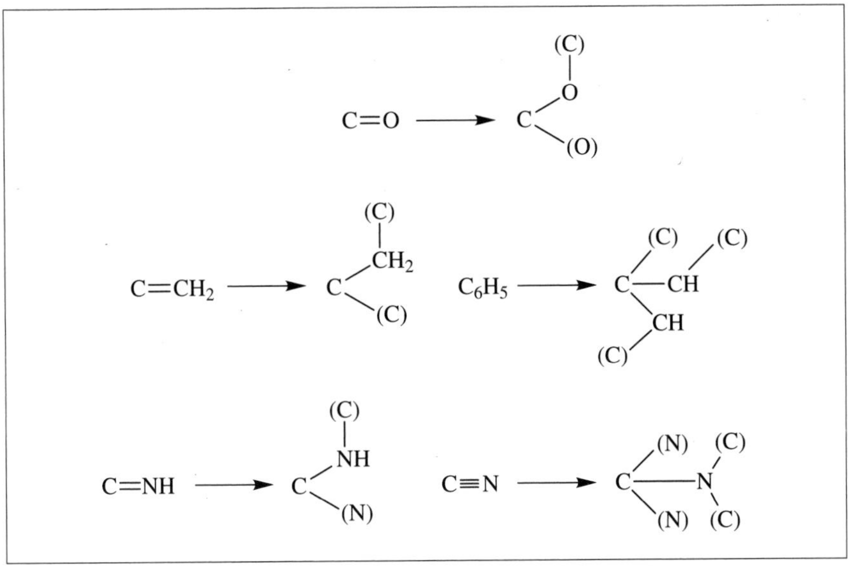

Figure 2.4

Table 2.1 shows the CIP priority sequences for a number of typical organic groups. A compound such as Cl$_3$C–CH(OH)–CH$_2$Br provides an instructive example on priorities. The group OH clearly has the highest priority, and H has the lowest. The decision between the other two groups cannot be made at the atoms bonded to the stereogenic carbon, as both these are carbon. However, it can be made on the basis of the substituents at these carbons; in particular, one looks for the single highest atomic number (and *not* the aggregate). Therefore, –CH$_2$Br takes precedence over –CCl$_3$.

For two isotopes, which of course have the same atomic number, priority is assigned on the basis of greater atomic mass, *i.e.* for the isotopes of hydrogen the priority is T > D > H. In the case of sulfoxides, for example, the lone pair is a formal substituent, and has the lowest priority of all.

Panico *et al.*[7] give a survey of the *R/S* convention.

2.3 Enantiomeric Excess, Enantiomeric Ratio

A sample that consists of one enantiomer of a chiral compound is said to be enantiomerically pure. This term has mainly replaced 'optically pure' that historically was used because the enantiomer content of a sam-

Table 2.1 CIP sequence rule order for commonly encountered groups (the lower the position of a group in the Table, the greater its priority)

1	H	hydrogen
2	D	deuterium
3	T	tritium
4	CH_3	methyl
5	CH_2CH_3	ethyl
6	$CH_2CH_2CH_3$	propyl
7	$CH_2CH_2CH_2CH_3$	butyl
8	$CH_2CH_2CH(CH_3)_2$	3-methylbutyl (isopentyl)
9	$CH_2CH(CH_3)_2$	2-methylpropyl (isobutyl)
10	$CH_2CH=CH_2$	prop-2-enyl (allyl)
11	$CH_2C(CH_3)_3$	2,2-dimethylpropyl (neopentyl)
12	$CH_2C\equiv CH$	prop-2-ynyl (propargyl)
13	$CH_2C_6H_5$	phenylmethyl (benzyl)
14	$CH(CH_3)_2$	1-methylethyl (isopropyl)
15	$CH=CH_2$	ethenyl (vinyl)
16	$CH(CH_3)CH_2CH_3$	1-methylpropyl (s-butyl)
17	C_6H_{11}	cyclohexyl
18	$CH=CHCH_3$	prop-1-enyl
19	$C(CH_3)_3$	1,1-dimethylethyl (t-butyl)
20	$(CH_3)C=CH_2$	1-methylethenyl (isopropenyl)
21	$C\equiv CH$	ethynyl
22	C_6H_5	phenyl
23	$4\text{-}CH_3C_6H_4$	p-tolyl
24	$4\text{-}NO_2C_6H_4$	p-nitrophenyl
25	$2\text{-}CH_3C_6H_4$	o-tolyl
26	CHO	methanoyl (formyl)
27	$COCH_3$	ethanoyl (acetyl)
28	COC_6H_5	benzoyl
29	$OCOCH_3$	methoxycarbonyl
30	$OCOC(CH_3)_3$	t-butoxycarbonyl
31	NH_2	amino
32	$^+NH_3$	ammonio
33	$NHCH_3$	methylamino
34	NHC_6H_5	phenylamino (anilino)
35	$NHCOCH_3$	acetylamino
36	$NHCOC_6H_5$	benzoylamino
37	$N(CH_3)_2$	dimethylamino
38	NO	nitroso
39	NO_2	nitro
40	OH	hydroxy
41	OCH_3	methoxy
42	$OCH_2C_6H_5$	benzyloxy
43	OC_6H_5	phenoxy
44	$OCOCH_3$	acetoxy
45	F	fluoro
46	SH	mercapto (sulfanyl)
47	SCH_3	methylsulfanyl (methylthio)
48	SO_2CH_3	methylsulfonyl
49	Cl	chloro
50	$SeCH_3$	methylselenanyl
51	Br	bromo
52	I	iodo

ple was always determined polarimetrically. Nowadays, other more accurate methods are employed in many instances (see Chapter 8).

The term 'enantiomeric excess', which gives a measure of the enantiomeric makeup of a sample that contains enantiomers A and B, is given by:

$$\% \text{ enantiomeric excess} = \frac{\text{no. of moles enantiomer A} - \text{no. of moles of enantiomer B}}{\text{no. of moles of both enantiomers}}$$

The terms in the numerator arise because what the polarimeter registers is a measure of the *difference* in population of the enantiomers in a sample. An enantiomeric excess of 50% signifies that the sample contains 75% of the enantiomer with, say, *R* configuration and 25% of *S*, that is 50% of a 1:1 mixture and 50% of the *R* enantiomer.

The enantiomeric content of a sample is expressed as an excess (*i.e.* a difference), or better as a ratio.

Since other techniques (*e.g.* NMR spectroscopy and gas–liquid chromatography) are used to derive enantiomer proportions under conditions in which each enantiomer gives its own 'signal' (Chapter 8), these are now more rationally expressed as a ratio rather than as an excess (this view was projected by Professor D. Seebach, ETH Zurich, at a colloquium at the University of Glasgow, 20 November 1998).

2.4   Racemization

A racemic mixture (or racemate) is defined as a 1:1 ratio of enantiomers. This corresponds to an enantiomeric excess of 0%, or an enantiomeric ratio of 1. Racemization involves the progressive loss of optical activity with time, usually in accord with a well-defined kinetic process.

A racemate is a 1:1 mixture of enantiomers.

In practice, spectra of racemic and enantiomeric forms are indistinguishable, though in principle this will only be true at infinite dilution. In the solid state the difference between a racemate and one enantiomer has features in common with a large box filled with (a) pairs of shoes and (b) left shoes only. Thus the packing in the two instances is not the same. This is reflected in different melting points and densities of the racemate and either enantiomer; these properties are the same for the two enantiomers.

Racemic compounds, or *dl* pairs, arise from (a) deliberate racemization of enantiomers by interconversion (the mechanism is not usually significant), (b) mixing the enantiomers in a 1:1 molar ratio, and (c) synthesis in the absence of a biasing influence that would cause one enantiomer to predominate.

It should be noted that on a molar basis a racemate is more stable than either enantiomer on account of the entropy of mixing (see Alberty and Silbey[8]). The entropy of mixing, $\Delta_{\text{mix}}S$, for the two enantiomers A and B present in equal amounts is given by equation (4):

$$\Delta_{mix}S = -nR(x_A \ln x_A + x_B \ln x_B) \qquad (4)$$

where n = total number of moles present, and x_A and x_B represent the mole fractions of A and B. The contribution of the entropy of mixing to the free energy of mixing is given by equation (5):

$$T\Delta_{mix}S = -nRT(x_A \ln x_A + x_B \ln x_B) \qquad (5)$$

For $n = 1$, and $x_A = x_B = 0.5$ and a room temperature of 298 K, the free energy of mixing has the value of $T\Delta_{mix}S = 1.71$ kJ mol^{-1}. This is the amount by which a mole of racemate is more stable than a mole of either enantiomer.

It should also be noted:

(i) that for a molecule with one stereogenic centre it is not possible to have a single molecule of a racemate; one needs at least both a molecule in which the stereogenic centre has R configuration and a molecule in which it has S (one can alternatively say that one needs both a molecule of D configuration and one of L configuration).

(ii) that a racemic sample of a compound is composed of chiral molecules, just as is the case for a sample of either enantiomer.

A term now in use to describe a sample that is partially racemized (or, alternatively, enantiomerically enriched), i.e. the sample consists of, say, 70% of enantiomer A and 30% of B, is scalemic (from the Greek 'scalemos' = lopsided), proposed by Brewster from Purdue University, Indiana (communication to Heathcock et al.[9]).

Separation of one or both enantiomers from a racemic mixture is called resolution, and it is the converse of racemization. Since resolution usually involves molecules or systems with more than one stereogenic centre, it is dealt with in Chapter 3.

2.5   Homochiral Molecules

The term homochiral was introduced by Kelvin in the 1904 publication of his Baltimore Lecture of 1884 and represented a relationship between molecules (see Mislow[10]). Molecules are homochiral if they possess the same sense of chirality. For example, the right hands of a group of people are homochiral (or alike). More recently, and unfortunately, homochiral has been used in the sense of enantiomerically pure, i.e. one reads of a 'homochiral compound', which clearly violates the original definition. Since some journals permit the latter usage, readers should be aware of the potential for confusion.

To illustrate this, the enzyme hog-kidney acylase hydrolyses the natural enantiomers of N-acylamino acids, regardless of the structure of the

R group in $RCH_2N(COMe)CO_2H$; *N*-acyl-L-amino acids are thus homochirally related (see Mislow and Bickart[11]).

Worked Problems

Q1. Assign configuration to compound **12**.

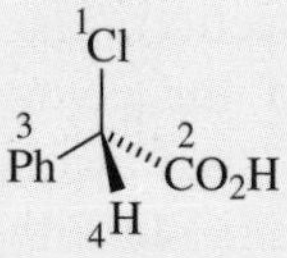

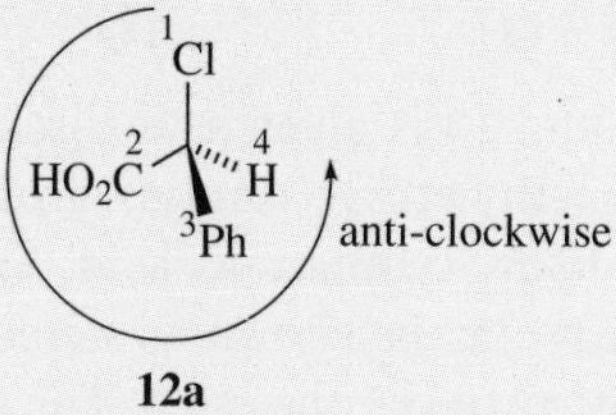

A. The stereogenic centre is C(2). Priorities of groups attached to the stereogenic centre are $Cl > CO_2H > Ph > H$. Rotate structure **12** around the C–Cl bond by 120° in an anti-clockwise direction. The resultant diagram is **12a**, and places the atom of lowest priority behind the page; as no bonds have been broken, there will be no change of configuration. With the eye looking toward, and along, the C(2)–H bond, the priorities decrease 1 > 2 > 3 in an anticlockwise sense and **12** therefore has *S* configuration. Compound **12** is (*S*)-2-chloro-2-phenylethanoic acid.

Q2. Assign configuration to compound **13**.

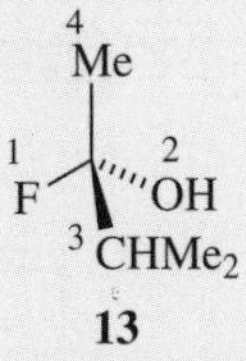

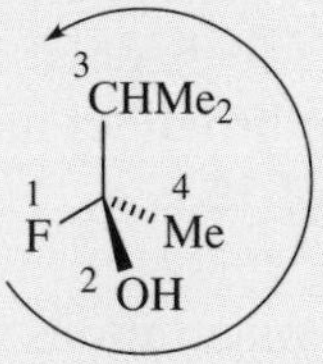

A. The stereogenic centre is C(2). Priorities of groups attached to the stereogenic centre are $F > OH > CHMe_2 > Me$. If one looks along the C(2)–Me bond from below the molecule, as drawn, one can see that the sequence of priorities decreases 1 > 2 > 3 in an anti-clockwise sense; hence the configuration is *S*.

Alternatively, rotate **13** clockwise around the C–F bond by 120°; this gives **13a**, which has the group of lowest priority behind the page. Again the configuration is *S*. Compound **13** is (*S*)-2-fluoro-3-methylbutan-2-ol.

Q3. Assign configuration to compound **14**.

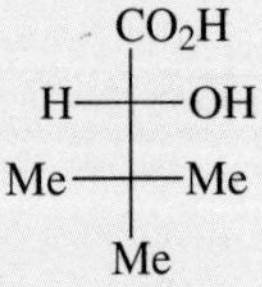

14 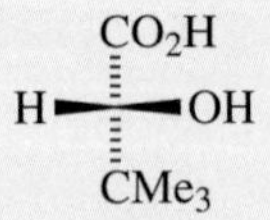 **14a** 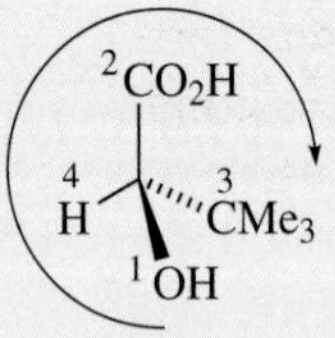 **14b**

A. Compound **14** is given as a Fischer projection. The lower part is a 1,1-dimethylethyl (*t*-butyl) group, and the compound, 3,3-dimethyl-2-hydroxybutanoic acid, may be re-drawn as in **14a**, which in turn may be shown as stereo diagram **14b** with C(1), C(2) and H in the plane of the page. The configuration is then assigned *R*.

Q4. Assign configuration to compound **15**.

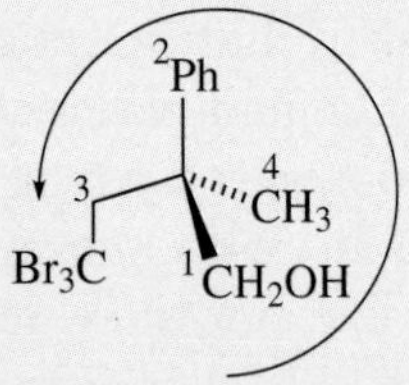

15

A. Priorities are assigned as shown. Note in particular that the CH_2–CBr_3 group is given priority 2, even though it contains three bromine atoms, behind CH_2OH and Ph. In this case the decision about priorities is made at the carbons bonded to the stereogenic centre, and the bromines in **15** are too remote to be decisive. The configuration of **15** is *S*.

Q5. Assign configurations to compound **16** (a hemiacetal), and its ethanoate (acetate) **17**.

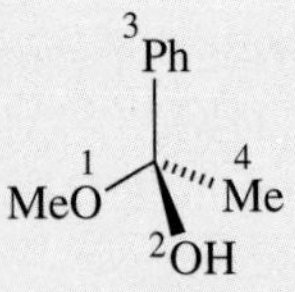

16 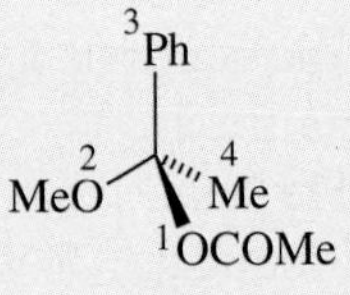 **17**

A. With priorities as indicated, **16** has *S* configuration. Although no bonds to the stereogenic centre are broken in conversion of **16** into the ethanoate **17**, the latter now has *R* configuration.

This pair of compounds illustrates a significant point. In assignment of priorities for compound **16**, OMe takes precedence over OH, whereas for **17**, OCOMe now has priority over OMe. Actually, the *absolute* configurations of **16** and **17** are the same, but they are correctly assigned different *R/S* symbols in the CIP convention.

Q6. Assign configuration to compound **18**, 1,1-dichloro-2-methyl-cyclohexane.

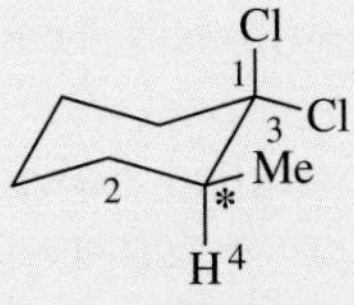

18

A. Cyclic compounds are treated in the same way as their open-chain analogues, and the priorities are assigned as shown. Note that the numbers in compound **18** refer to the CIP priority assignments around the stereogenic carbon C(2), which is denoted by C*. If, in accordance with the CIP rules, one looks at the C*–H bond from above the molecule as drawn, the configuration is seen to be *S*.

Summary of Key Points

- The principles of polarimetry, which are fundamental for the detection and characterization of chiral molecules, were outlined for molecules that contain a single stereogenic centre. Polarimetric measurements are most commonly carried out at 589.1 nm, the wavelength of the sodium D line. Specific rotation was defined.
- Two conventions that are very useful in defining the stereochemistry of molecules were described.
- The D/L convention is used for sugars and amino acids. Sugars are related to a standard, D-(+)-glyceraldehyde, which has one stereogenic centre.
- Naturally occurring amino acids are also related to L-(–)-glyceraldehyde by chemical transformations such as those indicated briefly for serine, which involves L-lactic acid as a relay.
- The *R/S* or CIP (Cahn–Ingold–Prelog) convention involves assignment of priorities, based on atomic numbers, to the four substituents at a stereogenic carbon or centre. This latter con-

vention is in almost universal use throughout organic chemistry.

- If a decision on the priorities cannot be made at the atoms bonded to the stereogenic centre, one proceeds further along the chain(s) until priorities can be assigned.
- Rules for the establishment of priorities were described for molecules that contain multiple bonds, such as in carbonyl and nitrile groups.
- A 50:50 mixture of enantiomers is called a racemate; a partially racemized compound, *e.g.* a mixture that contains, say, a 73:27 ratio of enantiomers is called scalemic. In this latter case the constitution of a mixture is described numerically by an enantiomeric excess or by an enantiomeric ratio.
- The term 'homochiral' was introduced.

Problems

2.1. Both the anti-inflammatory agent ibuprofen (**19**) and the antihypertensive agent methyldopa (**20**) have the *S* enantiomer as the active forms. Draw stereo diagrams of these active *S* enantiomers.

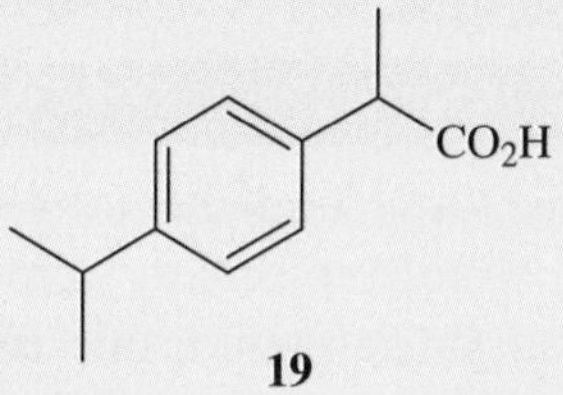

19 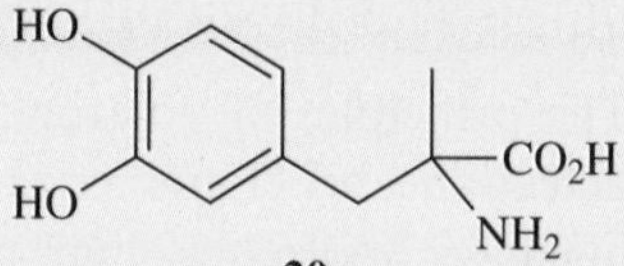   **20**

2.2. The potent anti-arthritic agent penicillamine (**21**) and the antinausea agent thalidomide (**22**) are drugs that have one stereogenic centre. For these compounds, one enantiomer [(*S*)-**21** and (*R*)-**22**, respectively] is beneficial and the other is highly toxic (**21**) or induces severe pre-natal deformities (**22**). Draw stereo diagrams of (*S*)-**21** and (*R*)-**22**.

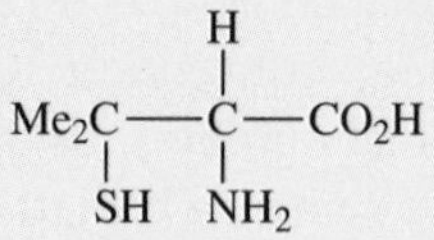

21 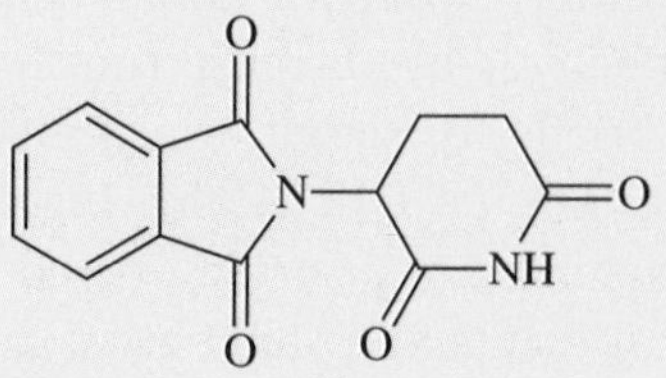   **22**

2.3. Draw stereo diagrams of:
(a) (S)-2-methylbutanoic acid, $EtCH(CO_2H)Me$;
(b) (R)-2-(ethoxycarbonyl)propanoic acid, $MeCH(CO_2Et)CO_2H$.

2.4. Assign R/S configurations to the stereogenic carbon in each of **23** and **24**.

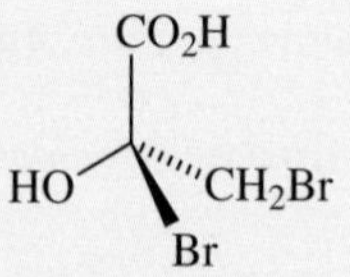

23

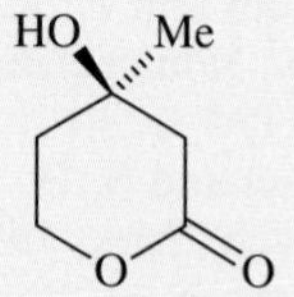

24

2.5. Convert the Fischer projections of **25–27** into stereo diagrams and then assign them R/S configurations.

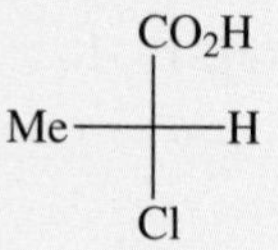

25

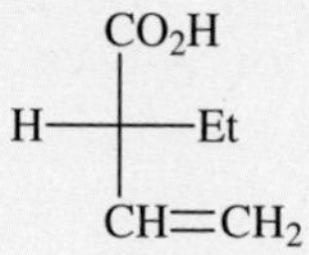

26

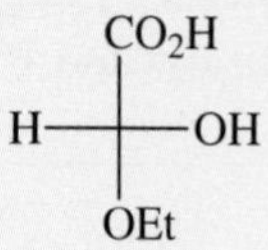

27

2.6. Assign R/S configurations to the naturally occurring amino acids **28–31**.

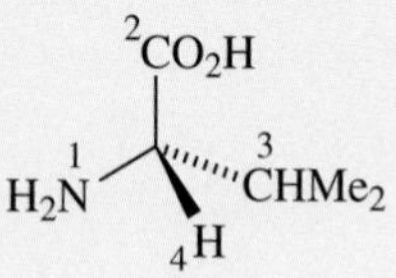

28 valine

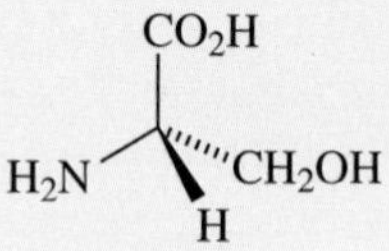

29 serine

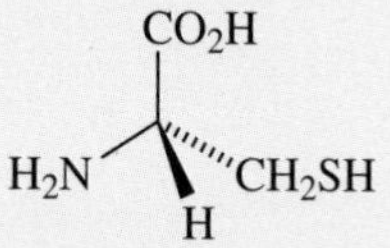

30 cysteine

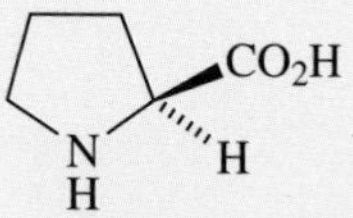

31 proline

2.7. Place the substituents in CIP priority sequence within each group:
(i) NO, NO_2; (ii) CH_2=$CHCH_2$, Me_3CCH_2; (iii) CHO, MeCO; (iv) CH_2=CH, Me_3CCH_2 and $PhCH_2$.

References

1. J. McMurry, *Organic Chemistry*, 4th edn., Brooks/Cole, New York, 1995, p. 294.
2. S. R. Buxton and S. M. Roberts, *Guide to Stereochemistry*, Longmans, Harlow, UK, 1996, p. 36.
3. R. S. Cahn, C. K. Ingold and V. Prelog, *Angew. Chem. Int. Ed. Engl.*, 1966, **5**, 385.
4. R. S. Cahn and C. K. Ingold, *J. Chem. Soc.*, 1951, 612.
5. Rules for Nomenclature of Organic Chemistry, *Pure Appl. Chem.*, 1976, **4**, 13.
6. E. L. Eliel and S. H. Wilen, *Stereochemistry of Organic Compounds*, Wiley, New York, 1994, p. 101.
7. R. Panico, W. H. Powell and J.-C. Richer (eds.), *Guide to IUPAC Nomenclature of Organic Compounds: Recommendations 1993*, Blackwell Scientific, Oxford, 1993, p. 149.
8. R. A. Alberty and R. J. Silbey, *Physical Chemistry*, 2nd edn., Wiley, New York, 1997, pp. 82 *et seq.*
9. C. H. Heathcock, B. L. Finkelstein, E. T. Jarvi, P. A. Radel and C. R. Hadley, *J. Org. Chem.*, 1988, **53**, 1922.
10. K. Mislow, *Topics Stereochem.*, 1999, **22**, 1.
11. K. Mislow and P. Bickart, *Isr. J. Chem.*, 1976/77, **15**, 1.

3

Molecules with Two (or More) Stereogenic Centres

Aims

By the end of this chapter you should be familiar with:

- Diastereoisomers in molecules with two or more stereogenic centres
- *meso* configurations when the substituents at two stereogenic centres are identical
- The terms *erythro* and *threo*, and *syn* and *anti*, when applied to molecules with two stereogenic centres
- Use of the terms 'epimers', *exo* and *endo* in connection with nomenclature of bicyclic molecules
- Separation of racemic compounds into enantiomers by resolutions that involve:
  Recrystallization of derived diastereoisomers
  Chiral chromatography
  Use of enzymes that react exclusively with one enantiomer

3.1　Enantiomers and Diastereoisomers

In the previous chapter it was shown that a molecule with one stereogenic centre gives rise to two enantiomers. A molecule with two stereogenic centres gives rise to a *maximum* of four stereoisomers; a molecule with three stereogenic centres can have a maximum of eight stereoisomers, and so on. In general, if a molecule contains n stereogenic centres, a maximum of 2^n stereoisomers can exist.

A molecule with n stereogenic centres can have a maximum of $2n$ stereoisomers.

　The molecule **1** has two stereogenic centres and has a different set of substituents at each centre; accordingly, there are four stereoisomers of **1**. Four stereoisomers also result if in **1** one replaces, say, an NH_2 by an

O
‖
MeC＼
    O
F—C—C—NH₂   Ph
Br   OMe
       1

F, so that one substituent, *i.e.* F, is now common to both stereogenic centres.

In designating the stereochemistry of, say, **1**, the molecule is first numbered systematically. The configuration at each stereogenic centre is determined in the same way as for molecules with one stereogenic centre (Chapter 2). The symbol *R* or *S* is then associated with a particular carbon by its number and the pair of symbols is placed in brackets. Accordingly for **1** the four stereoisomers can be (1*R*,2*R*), (1*R*,2*S*), (1*S*,2*S*) and (1*S*,2*R*), and these are shown schematically in Figure 3.1.

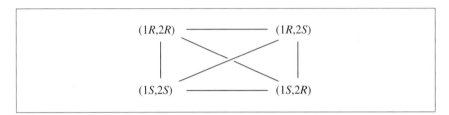

Figure 3.1

It can be seen that the stereoisomers with configuration (1*R*,2*R*) and (1*S*,2*S*) are related as enantiomers, and the (1*R*,2*S*) and (1*S*,2*R*) stereoisomers are also related as enantiomers. What then is the relationship between the stereoisomers of **1** with configurations (1*R*,2*S*) and (1*R*,2*R*), for example? With respect to each other, these have one stereogenic centre of common, (1*R*), configuration, and the other of opposite, (2*S*) and (2*R*), configuration. Clearly these stereoisomers are not enantiomers. Instead, they are related as **diastereoisomers** and the same can be said for the (1*S*,2*R*) and (1*S*,2*S*) stereoisomers. The most concise definition, given in the plural, is: 'diastereoisomers are stereoisomers that are not enantiomers'.

Diastereoisomers have different physical properties and react at different rates: they are stereoisomers that are not enantiomers.

For a compound that has two stereogenic centres, and has different groups attached to each such centre as in **1**, the representation in Figure 1 shows that molecules related vertically are enantiomers, whereas those related either diagonally or horizontally are diastereoisomers. The interrelationships in Figure 1 show that whereas a given chiral molecule can possess only one enantiomer, it can have more than one diastereoisomer.

A particular stereoisomer of **1** with two stereogenic centres can have only one enantiomer, but more than one diastereoisomer.

Molecules that are related as diastereoisomers have different melting points and solubilities, and have different rates of reaction. In some cases, these different rates of reaction can lead to a particular product being formed from a reaction of one diastereoisomer but not the other. Use is made of differences of, for example, solubility of diastereoisomers in the resolution of a racemic mixture (Section 3.6).

3.2 *Meso* Configuration

In contrast to the general case of **1**, we now consider molecules with two stereogenic centres, but which carry an identical set of substituents at each carbon. We then analyse the relative configurations of these substituents and the stereochemical consequences.

As a prototype molecule, tartaric acid (**2**; 2,3-dihydroxybutanedioic acid), shown here without stereochemical specification, is chosen. Tartaric acid is a compound of considerable historical importance (Section 3.6), and derivatives of tartaric acid are still compounds of contemporary research interest (see Seebach *et al.*[1]). (2*R*,3*R*)-(+)-Tartaric acid is shown in sawhorse projection in **3**.

A molecule with two stereogenic centres, each of which carries the same substituents (C_{abc}–C_{abc}), *e.g.* tartaric acid, is limited to a total of three rather than four stereoisomers.

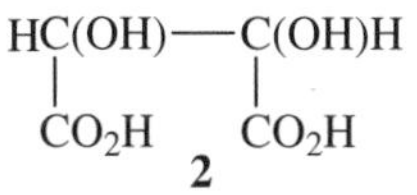

$$
\begin{array}{cc}
\text{HC(OH)} & \text{C(OH)H} \\
| & | \\
\text{CO}_2\text{H} & \text{CO}_2\text{H} \\
\end{array}
$$

2

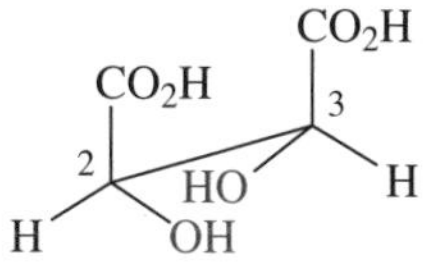

3

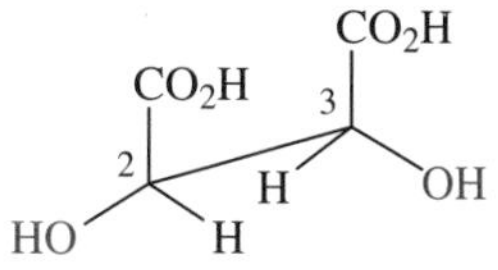

4

> With the aid of molecular models verify that the configurations of C(2) and C(3) are correct.

The enantiomer of **3** is shown in **4** and is (2*S*,3*S*)-(−)-tartaric acid. Of course, in principle any conformation of **3** may be used, with its mirror image, to demonstrate that the compounds in the pair are enantiomers; however, use of the eclipsed forms **3** and **4** is visually easier.

Fischer projections of **3** and **4** are given in **5** and **6**, respectively, and the corresponding Newman projections, looking along the C(2)–C(3) bond, are shown in **7** and **8**, but now with the rotation of the 'back' groups around the C(2)–C(3) bond such that the molecules are shown in staggered and visually less cramped conformations.

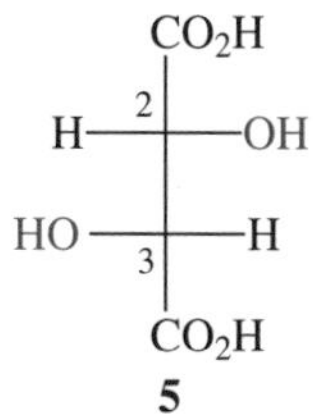

5

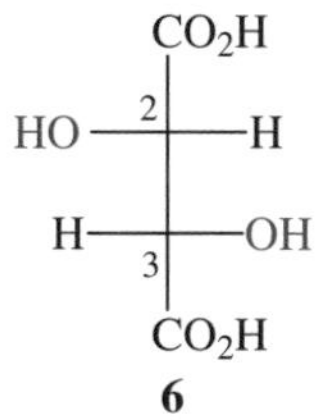

6

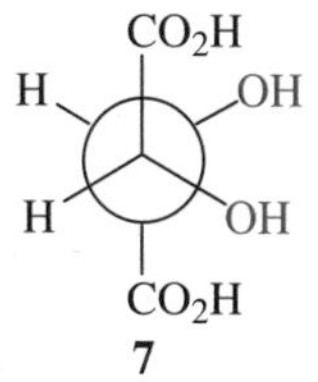

7

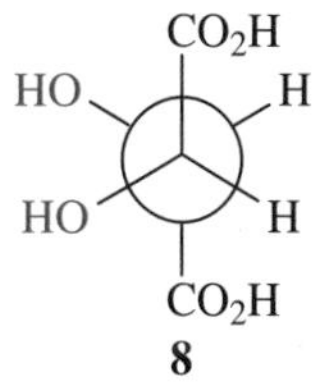

8

We now consider the other configurations of tartaric acid, *i.e.* (2*R*,3*S*) **9** and (2*S*,3*R*) **10**. Here, the presence of identical sets of substituents at C(2) and C(3) brings about a difference from the general case which was shown by **1**. Inspection of **9** and **10** reveals that **10** is a mirror image of **9** (and *vice versa*). However, **9** and **10** are identical and superimposable.

The *meso* stereoisomer of tartaric acid is achiral, and possesses two self-cancelling stereogenic centres of opposite configuration.

CO$_2$H
CO$_2$H   3
2   H   OH
H   OH

CO$_2$H
CO$_2$H   3
2   HO   H
HO   H

9   **10**

> Use molecular models to check that this is so.

Therefore, and importantly, this molecule is achiral despite containing two stereogenic centres. This can be shown in two ways.

(a) If one focuses on **9**, in the conformation shown, there is seen to be a plane of symmetry in this molecule half-way along the C(2)–C(3) bond, such that –CO$_2$H reflects on to –CO$_2$H, –OH on to –OH, and –H on to –H.

(b) Take a molecular model of **9**, and break the C(2)–C(3) bond in half, and then mark the 'break' on both sides with tape, or similar, and regard the taped ends as identical substituents. Examination of the resultant taped fragments clearly reveals that C(2) and C(3) are of opposite configuration.

In Chapter 2, a number of everyday objects were considered; some are chiral and some are not. Those that are not chiral have a plane of symmetry. The same characteristics operate at a molecular level and one can therefore state that if a molecule possesses a plane of symmetry it is not chiral. It is possible to see this for simple molecules such as methane and dichloromethane and also for *meso*-tartaric acid **9**, drawn so that the plane of symmetry may readily be recognized. In assessment of whether a molecule is, or is not, chiral it is advisable to check for both (a) the presence of stereogenic centres and (b) the presence of a plane of symmetry.

> As an exercise, take a model of **9**, as drawn, and rotate the rear groups clockwise by 60° around the C(2)–C(3) bond. Then confirm that in this conformation the molecule is now chiral.

However, **9** is *not* chiral and one can accordingly amplify the definition above to state: if a molecule possesses a plane of symmetry *in any energetically available conformation*, it cannot be chiral. This statement holds whatever the complexity of the molecule. However, it should be noted that chiral molecules can possess axes of symmetry, and such molecules are considered in Chapter 5.

The achiral nature of **9** is also implicit in the designation of configurations (2*R*,3*S*) or (2*S*,3*R*) with identical sets of substituents at the stereogenic centres. As regards optical activity, the effects of C(2) and C(3) in **9** are self-cancelling because the stereogenic centres have opposite configuration; indeed, in early literature compounds such as **9** were described as 'internally compensated'. Nowadays, such compounds are given the prefix *meso*, and use of the term is general; *meso* compounds are always achiral (and optically inactive) even though they have two or more stereogenic carbons.

> Confirm that **3** and **4** are enantiomers, and that **3** and **9**, also **4** and **9**, are diastereoisomers.

3.3　*Erythro/Threo* and *Syn/Anti* Configurations

Let us now look at molecules with two stereogenic centres, usually adjacent, which have two common substituents. The names of the two configurations are taken from two carbohydrates, but are now used quite generally. D-Erythrose shown in Fischer **11** and sawhorse **12** projections in its open-chain forms is such a compound. In **12** the pairs of like substituents are shown in an eclipsed conformation.

A molecule with two stereogenic centres with two substituents in common (C_{abc}–C_{abd}) gives rise to erythro *and* threo *diastereoisomers.*

Firstly, one might ask: why is **11** given the configuration D? This assignment is based on the configuration of the stereogenic carbon that is furthest from the carbonyl group [*i.e.* C(3) in 11] with respect to that of D-glyceraldehyde (**13**). Since they are the same, **11** is given D configuration. This protocol can also be applied to longer chain sugars such as pentoses and hexoses (erythrose and threose are themselves tetroses). If, as in **12**, it is possible to eclipse each of the like substituents, H with H, OH with OH, and also the 'unlike' substituents CHO and CH$_2$OH, the configuration is *erythro*. The nomenclature is quite general; thus *erythro*-3-phenylbutan-2-ol is shown in **14**. Of course, **12** and **14**, and indeed all *erythro* compounds, have enantiomers, but this does not concern us here.

The second configuration is termed *threo*, and originates from the sugar 'threose'; D-threose is shown in Fischer **15** and sawhorse **16** projections. With respect to D-erythrose (**11**), only the configuration at C(2)

13   **14**

has changed. Two consequences follow from this: (i) **11** and **15** are related as diastereoisomers and (ii) the stereochemical symbol is D in both **11** and **15**.

15   **16**   **17**

If a molecule (C_{abc}–C_{abd}) in an *erythro* configuration is converted into one in which the unlike substituents, say c and d, are each transformed into a common substituent, say e, *i.e.* to give (C_{abe}–C_{abe}), the resultant compound is *meso*.

The configurations at C(2) and C(3) in D-threose are now such that in sawhorse projections one can eclipse only (a) the H atoms, or (b) the OH groups or (c) the unlike substituents, CH_2OH and CHO, at any one time. Again the nomenclature is quite general, and one enantiomer of *threo*-3-phenylbutan-2-ol is shown in **17**.

An *aide-memoire* to assist in assignment of *erythro* and *threo* configurations is this: if one takes erythrose (**11**), the product obtained after oxidation of both the CH_2OH and CHO groups to COOH is *meso*-tartaric acid (**9**).

Use of the terms *erythro* and *threo* is quite explicit in the context of two adjacent stereogenic centres, each of which has two sets of substituents in common. However, some confusion has arisen when the *erythro/threo* nomenclature has been applied to systems in which there is only one common substituent on the two adjacent carbons. Such a case is found in the products of asymmetric aldol reactions. The mechanism of this reaction lies outside the scope of this chapter; however, we give here the basis of an alternative nomenclature system, which is now becoming more common for products of this reaction, and which was introduced by Masamune et al.[2] The main chain is drawn as a zig-zag, and lies in the plane of the paper. Substituents that lie on the same side of the plane are termed *syn*, and those that do not are termed *anti*. Thus in the case of **18** the 3,4 stereochemistry is *syn* and in **19** it is *anti*; **18** can be fully described as *syn,syn* and **19** as *syn,anti*. The drawings here represent essentially a telescoping of sawhorse projections, and the position of the eye has moved so as to view the molecule from a more 'sideways on' aspect. The wedges imply that the bond comes out from the paper toward the viewer and the 'hatched' (or broken) lines imply that the bond is directed behind the paper. Although sawhorse projections are ideal for inspection of the configurations around a two-carbon unit,

A *syn/anti* notation for relative configuration at sp³ carbons is described.

Loan Receipt
Liverpool John Moores University
Learning and Information Services

Borrower ID:	21111130102114
Loan Date:	17/07/2008
Loan Time:	3:26 pm

Stereochemistry /
31111009304146

Due Date: 19/09/2008 23:59

Stereochemistry /
31111011467097
Due Date: 19/09/2008 23:59

Organic chemistry /
31111010719167
Due Date: 07/08/2008 23:59

Please keep your receipt
in case of dispute

Boon' Shing

was

here !

they become unwieldy when chains of more than two carbons are involved, and in such cases stereo drawings are now preferred. A saw-horse projection of **19** along the C(4)–C(3) bond is given in **20** (R' = cyclohexyl).

As an alternative to *erythro* and *threo*, two further conventions have been introduced for the specification of relative configuration in molecules that contain two (or more) stereogenic centres. In a molecule such as (2*R*,3*R*)-tartaric acid (**3**) or (2*S*,3*S*)-tartaric acid (**4**), the configurations are described as *l* for like. *meso*-Tartaric acid (**9**) has configuration (2*R*,3*S*) [or (2*S*,3*R*)] and is described as *u* for unlike. Compounds with more than two stereogenic centres, *e.g.* (1*R*,2*R*,3*S*), have relative configurations specified by a comparison of their configurations in sequence. A compound with the above configuration, and its enantiomer, can be collectively described as *l.u.* (see Seebach and Prelog[3]).

Alternatively, in a convention employed by *Chemical Abstracts*, the two enantiomers (2*R*,3*R*)- and (2*S*,3*S*)-tartaric acid can be specified by (2*R**,3*R**). The symbols (2*R**,3*R**) indicate that if C(2) has *R* configuration, so does C(3), and likewise that if C(2) has *S* configuration, so does C(3). *meso*-Tartaric acid can formally be described as having (2*R**,3*S**) configuration and this indicates that if C(2) has *R* configuration, C(3) has the *S* and *vice versa*.

Note that whereas *R* and *S* symbols apply to specific carbons, D and L apply to particular *reference* carbons, as has been described for sugars in this section. In the case of amino acids the configuration, D or L, is always taken from the α carbon, which is bonded to both NH_2 and CO_2H. This applies even when there is more than one stereogenic centre in the amino acid, as in the case of, for example, threonine. Further discussion of this point lies outside the scope of this text but is dealt with explicitly in the text by Loudon.[4]

3.4   Caged Compounds with Two Stereogenic Bridgehead Carbons

An example of a compound with two stereogenic carbons in a rigid caged structure is camphor, a naturally occurring bicyclic ketone which is more abundant in the (+)-form **21**. The presence of the carbonyl group in camphor results in two stereogenic centres, at C(1) and C(4).

A bicyclic compound, chiral because of stereogenic bridgehead carbons, is restricted to two enantiomers, and no diastereoisomers.

Verify that **21** has the configuration (1*R*,4*R*).

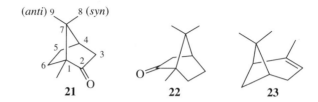

21   **22**   **23**

Accordingly, in **21**, C(1) has *R* configuration, as has C(4). It is not possible to construct a camphor in which C(1) has *R* configuration and C(4) has *S* configuration. The (+)-camphor shown in **21** is (1*R*,4*R*)-camphor, but on account of the cage structure it is sufficient to specify **21** as (1*R*)-camphor. The only other possibility is the more expensive enantiomer **22**, which is (1*S*,4*S*)-camphor, and this enantiomer is often referred to as (1*S*)-camphor. Camphor and other chiral caged molecules with stereogenic centres at bridgehead carbons, *e.g.* α-pinene (**23**), which is extracted from pine trees, are examples of molecules that possess two stereogenic centres, but which have only two stereoisomers, related as enantiomers of each other. Of course, derivatives of such molecules may contain additional stereogenic centres. If there are additional stereogenic centres, these are independent of the restriction on the number of stereoisomers that may exist, which is imposed by the bridgehead stereogenic centres.

3.5 Epimers, and the Nomenclature of Bicyclic Compounds

Epimers are diastereoisomers whose configurations differ at only one carbon.

The term epimer originated in sugar chemistry and can be illustrated with the examples of D-lyxose (**24**) and D-xylose (**25**). There are three stereogenic centres in each of these molecules; the configurations at C(3) and C(4) are the same in both, whereas the configuration of C(2) in **24** differs from that in **25**.

Two diastereoisomers that differ in configuration at only one stereogenic carbon are called **epimers**. The term is quite general, though it is

$$^1\text{CHO}$$
$$\text{HO}\overset{2}{-}\!\!-\!\!\text{H}$$
$$\text{HO}\overset{3}{-}\!\!-\!\!\text{H}$$
$$\text{H}\overset{4}{-}\!\!-\!\!\text{OH}$$
$$^5\text{CH}_2\text{OH}$$

24

$$\text{CHO}$$
$$\text{H}\!\!-\!\!-\!\!\text{OH}$$
$$\text{HO}\!\!-\!\!-\!\!\text{H}$$
$$\text{H}\!\!-\!\!-\!\!\text{OH}$$
$$\text{CH}_2\text{OH}$$

25

rarely applied to molecules with only two stereogenic centres. Consider, for example, reduction of (1*R*)-camphor (**21**) with LiAlH$_4$ to give two alcohols, isoborneol (**26**) and borneol (**27**). These alcohols fulfil the criteria for being epimers; their configurations are the same at C(1) and C(4), and are opposite at the hydroxyl-bearing carbon, C(2). Although the configurations at C(2) in **26** and **27** can, of course, be specified by the usual *R,S* convention, there is an alternative and convenient way to describe them. This uses the terms *exo* and *endo*, and makes use of the number of ring carbons that constitute the bridges which connect the bridgehead carbons at C(1) and C(4). A substituent on a particular bridge is called *exo* if it is on the *same side* as the smaller bridge, *i.e.* the one-carbon bridge formed by C(7) [and in which C(7) is bonded to two methyl groups]. On the other hand, the OH group in **27** is described as *endo* because it is *opposite* the smaller bridge. In reduction of **21**, the preferred approach of LiAlH$_4$ to the carbonyl group is from the less-hindered *endo* direction with the resultant formation of the *exo* alcohol **26** as the major product. The designations *endo* and *exo* thus serve as internal markers only, *i.e.* in specifying that the OH group is *syn* (same side as) or *anti* (opposite) to the shorter bridge. These terms do not specify whether the configurations of the OH-bearing carbons in **26** and **27** are *R* or *S*.

The systematic naming of bicyclic compounds is now considered with norbornane (**28**), a C$_7$H$_{12}$ compound, used to demonstrate the protocol, which is due to the German chemist von Baeyer. Firstly, the bridgehead carbons C(1) and C(4) are marked for reference. Secondly, the number of carbons between the bridgehead carbons in each of the three bridges is counted, and the molecule positioned so that the number in the right (*r*), left (*l*) and top (*t*) bridges are in the sequence $r \geq l \geq t$ and placed in brackets. Compound **28** is thus known as bicyclo[2.2.1]heptane.

The sum of the numbers inside the bracket is two less than the total number of carbons because the two bridgehead carbons C(1) and C(4) are not included. The term 'bicyclo' is used because although **28** contains a total of three rings, any two rings embrace all ring carbons; however, with this approach, ambiguity can arise in compounds with more than two rings. More rigorously, the term 'bicyclo' in **28** is used because the compound has two double bond equivalents; that is its molecular formula (C$_7$H$_{12}$) corresponds to two pairs of hydrogens fewer than heptane, C$_7$H$_{16}$.

Four further examples are given here. Compound **29**, C$_8$H$_{14}$, is bicyclo[3.2.1]octane; compound **30** (X = H) is bicyclo[2.2.2]octane; and in the case of **31** (X = OH) the hydroxyl group is neither *endo* nor *exo* as both the unsubstituted bridges contain equal numbers of carbons; the molecule is, however, chiral solely because of the stereogenic centre at C(2), the hydroxyl-bearing carbon. Compound **32**, whether the ring junction is *cis* or *trans* (Section 6.6), is bicyclo[4.4.0]decane, though it is more

An *exo* epimer has a substituent directed toward the smallest bridge; substituents directed toward the largest bridge are *endo*.

26 **27**

28

The nomenclature of, *e.g.* bicyclo[2.2.1]heptane and analogues, is described.

commonly known as decalin. The case of an additional ring is exemplifed by the C_7H_{10} compound nortricyclene (**33**). The systematic name for **33** is tricyclo[2.2.1.0^{2,6}]heptane, the last entry within the bracket signifies that, compared to **28**, an additional zero-carbon bridge has been introduced between C(2) and C(6) to form the three-membered ring.

29   **30** X = H   **32**   **33**
          31 X = OH

There are a couple of further points, one of which is exemplified by camphor (**21**) and the borneols. These molecules have the functional groups on one of two two-carbon bridges. Accordingly, the numbering is arranged to be as low as possible, and so in camphor, for example, the carbonyl carbon is C(2).

If a substituent on the smallest bridge is directed toward the next smallest bridge, it is termed *syn*; if the substituent points toward the largest bridge, it is termed *anti*. However, camphor, for example, has two bridges that both contain two carbons, *viz.* C(2)–C(3) and C(5)–C(6). Now if the substituent on the smallest bridge points toward the bridge that itself carries substituents [in camphor C(2) is part of the carbonyl group], it is called *syn*. Where the substituent on the smallest bridge in camphor points toward the unsubstituted bridge, it is called *anti*. In camphor (**21**) the methyl groups containing C(8) and C(9) are therefore labelled *syn* and *anti*, respectively.

3.6   Separation of Enantiomers: Resolution

Resolution can be thought of as the converse of racemization (Section 2.4). One starts with a 50:50 mixture of both enantiomers and separates this mixture into the individual enantiomers. Of course, for some purposes one may only want one enantiomer, and recovery of the second enantiomer can be painstaking. Since enantiomers have identical properties, including solubility, separation of enantiomers by recrystallization is quite rare. It was, however, such a crystallization by Pasteur in 1848 that opened up the field of resolution. Pasteur's key observation was that two distinct but related types of crystal were obtained from an aqueous solution of the sodium ammonium salt of racemic tartaric acid. The two types of crystal were related as object and non-superimposable mirror image, and one type was identical to the dextrorotatory crystals of sodium ammonium tartrate obtained from (+)-tartaric acid, itself obtained as a by-product of wine-making.

Pasteur separated his crystals manually with the aid of a magnifying glass and tweezers. Moreover, he demonstrated that solutions of the two types of crystals had the same value of specific rotation, though of opposite sign. In correlating the macroscopic with the molecular, the important and fundamental conclusion was drawn that the molecules from which the two crystals were composed were likewise non-superimposable mirror images. Like many important discoveries, Pasteur's work was a combination of perceptiveness, tenacity and happenstance, the last because he worked with the only known tartrate salt that gives enantiomeric crystals with different crystal forms. Additionally, these different crystals form only at temperatures below 25 °C. From a knowledge that he worked in Paris, one can conclude that Pasteur did not carry out his experiments at the height of summer.

Typically, resolution depends on the conversion of enantiomers, which possess identical physical properties, into diastereoisomers, which do not, and then exploiting the difference in physical properties in order to separate the diastereoisomers. Finally, the diastereoisomers are reconverted to the component, and now separated, enantiomers.

In schematic terms the separation of racemic compound R,S into enantiomers by reaction with a single enantiomer of R^* is represented in Scheme 3.1. RR^* and SR^* are diastereomers, and can be covalent compounds or salts. R^* is chosen so that the solubilities of RR^* and SR^* in a particular solvent are different. The racemic mixture, now in combination with R^*, $i.e.$ RR^* and SR^*, is now recrystallized and one diastereoisomer, say RR^*, will crystallize out whereas the other, SR^*, will stay in solution. Sometimes when the first crop of crystals is merely enriched in RR^*, a number of further recrystallizations will be necessary. Salts are the preferred diastereoisomers (RR^* and SR^*), because they are both easily formed and readily re-converted to the starting enantiomers, following separation.

> Resolution by recrystallization depends on conversion of enantiomers into diastereoisomers, usually salts, and then exploiting solubility differences.

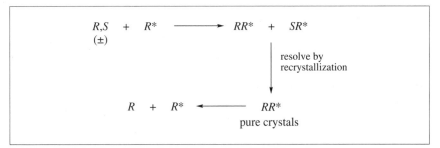

Scheme 3.1 Resoloution of a racemic compound R,S to give the more insoluble RR^* as crystals, whereas the diastereoisomer SR^* remains in solution

Resolution of a carboxylic acid can be achieved directly by formation of diastereomeric salts with an amine. Naturally occurring bases such as brucine (**34**) or ephedrine (**35**) are often used, though it should be mentioned that choice of a suitable amine is usually a matter of trial and

Resolution of alcohols usually involves synthesis of a phthalate half-ester, then formation of a diastereomeric salt.

error and is probably related to crystal packing of the diastereoisomeric salts. In common with camphor, brucine, a naturally occurring alkaloid, has its stereogenic centres in fused ring systems; nature makes only one enantiomer, (–)-brucine (**34**).

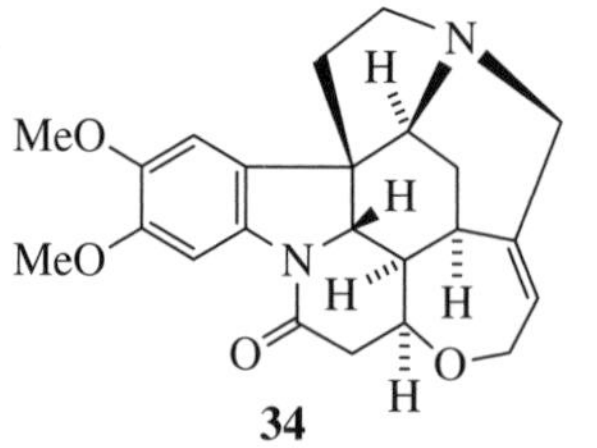

34

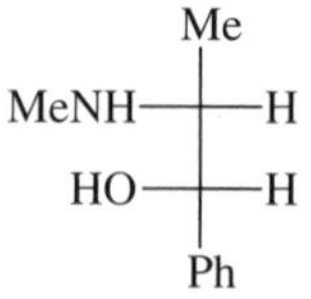

35

> How many stereogenic centres are there in brucine?

Resolution of alcohols requires a more ingenious approach in order to acquire suitable diastereomeric salts for resolution by recrystallization. The racemic alcohol, ROH, to be resolved is first converted to its phthalate monoester **36** (phthalic acid is benzene-1,2-dicarboxylic acid). The strategy here is to leave one carboxylic acid group free, and available to form diastereoisomeric salts with an amine such as **34** or **35**. Once the appropriate ammonium salt of **36** has been resolved, the amine is removed by acidification to give one enantiomer of **36**, and this ester is then hydrolysed with alkali to give one enantiomer of ROH. The more soluble diastereomeric ammonium salt can then, in principle, be processed similarly to yield the other enantiomer.

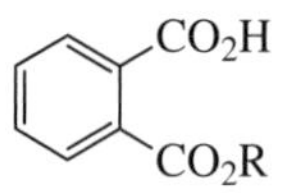

36

3.7 Separation of Enantiomers by Chromatography

The success of chiral column chromatography depends on diastereoisomeric interactions between the chiral stationary phase and the enantiomers; this leads to differential adsorption of enantiomers.

It is also possible to resolve enantiomers by means of column chromatography. If a column consists of, or contains, a suitable chiral stationary phase, the enantiomers should be eluted, *i.e.* washed from the column, at different rates. This is because, although there is no formation of chemical bonds, the enantiomers now experience diastereomeric interactions with the chiral stationary phase. If there is a high degree of 'chiral recognition' of one particular enantiomer by the stationary phase, then it will be preferentially adsorbed on the chiral stationary phase and accordingly eluted more slowly. Good separations of quite sizeable quantities (50 g per elution) can be achieved in some cases. Column chromatography is described in the books by Harwood *et al.*[5] and Armarego and Perrin;[6] chiral chromatography is the subject of a paper by Pirkle *et al.*;[7] see also Lochmüller.[8]

Chiral stationary phases can consist of starch, which for instance allows almost complete resolution of mandelic acid (2-hydroxy-2-phenylethanoic acid), $PhCH(OH)CO_2H$. Synthetic stationary phases such as **37**, derived from the enantiomerically pure amino acid alanine, and **38**, likewise from valine, are effective in resolution of alcohols, amines and amino acids (both α and β), with the proviso that these compounds first be derivatized, *e.g.* an alcohol must be transformed into an ester before being passed down the column.

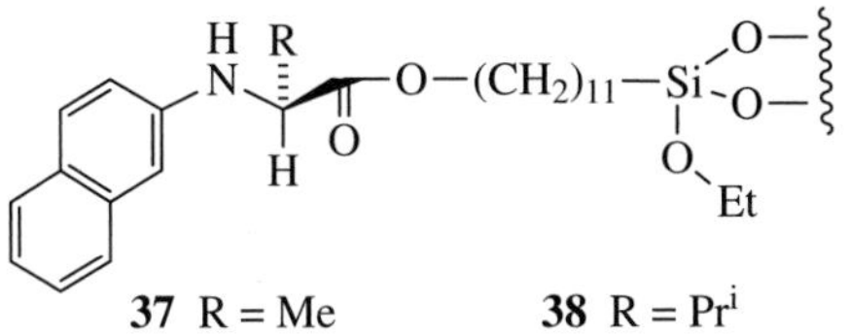

37 R = Me   **38** R = Pr^i

3.8   Resolution with Enzymes

Enzymes are compounds of high molar mass, generally contain many amide groups, and catalyse certain reactions in living cells. All enzymes are catalysts; they are themselves chiral, and in their natural aqueous environment are very stereoselective. Some enzymes are highly selective in the reactions that they catalyse, whereas others are more broadly receptive, and in general enzyme-catalysed reactions tend to be rapid. A discussion of the characteristics of enzymes and their mode of action is given by Stryer.[9] Furthermore, enzymes have been instrumental in bringing about resolution of enantiomers under laboratory conditions. A quantitative analysis of biochemical kinetic resolutions of enantiomers has been made by Chen *et al.*[10]

In order to give a flavour of the method, three examples are presented here.

As part of a project to synthesize one enantiomer of a chiral insect pheromone, resolution of **39** was required. This was achieved by enantioselective hydrolysis of racemic **39** to produce alcohol **40** (Scheme 3.2).

A suitable enzyme catalyses enantioselective hydrolysis of ester **39**. This gives an alcohol of one configuration and leaves behind the ester of the other. Differentiation between enantiomers by enzymes is almost total.

Verify that this alcohol has *R* configuration.

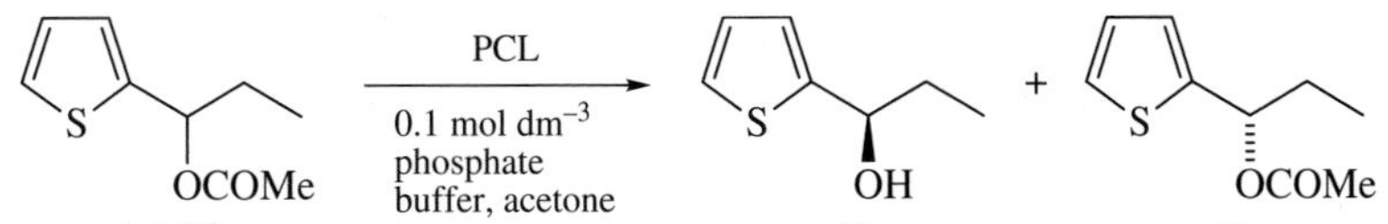

Scheme 3.2

The other enantiomer was left untouched by *Pseudomonas cepacia* lipase (PCL), the enzyme of choice. After reaction is complete, the resolved alcohol must be separated from its enantiomeric acetate **41**. It is noteworthy that the enzyme functions in a buffered aqueous acetone medium; acetone does not interfere with the action of the enzyme.

There is now a sufficient data bank of enzymic resolutions for these to guide the choice of enzyme for a new resolution. Although at 18 °C and after 60% conversion it was possible to obtain acetate **41** of 99% enantiomeric excess (*ee*), it should be recognized that use of enzymes is not always the complete answer. In the above resolution, smaller *ee* values were obtained at higher temperatures and at larger conversions, findings that are perhaps not surprising if one considers that the enzymes are here operating outside their natural environment (see Kang *et al.*[11]).

In a second investigation, racemic 2-hydroxy-4-phenylbutanoic acid (**42**) was reacted with lipase PS (LPS) and vinyl acetate (VA) (ethenyl ethanoate) in *t*-butyl methyl ether (2-methoxy-2-methylpropane) to give acetate (*S*)-**43** in 35% yield and in 99% *ee*; unreacted alcohol **44** was found to have 99% *ee* with *R* configuration predominant (Scheme 3.3).

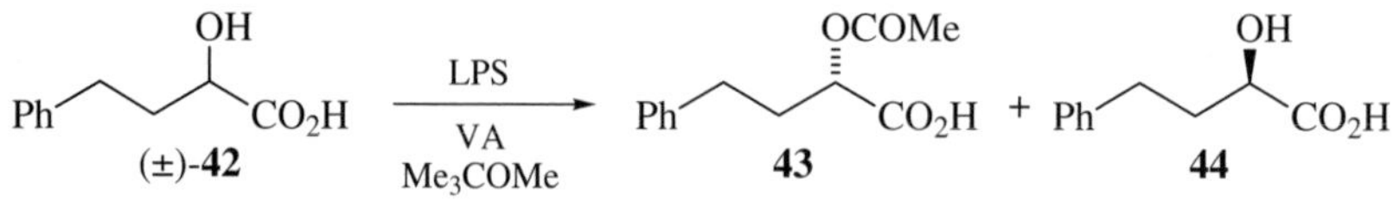

In the presence of vinyl acetate as acylating agent, racemic hydroxy acid **42** was converted enzymatically into the (S)-acetate; the (R)-hydroxy acid was unaffected.

Scheme 3.3

In this resolution involving transesterification, the enzyme is catalysing the acetylation by vinyl acetate of the hydroxyl group of **42** with *S* configuration. This use of vinyl acetate is now fairly common in enzymatic resolution (see Chadha and Manohar[12]); hexanoic anhydride is also sometimes used in this context.

Thirdly, the lipoprotein lipase from *Psuedomonos* species (PSL) reacted with racemic cyanohydrin **45** to give the acetate (ethanoate) **46** of *S* configuration with 98% *ee* after 59% conversion (Scheme 3.4). The unreacted alcohol, effectively now resolved into the enantiomer of *R* configuration **47**, because of its rejection by the enzyme, was then converted into the hydroxy ester **48**, which is an important intermediate in the synthesis of an angiotensin-converting enzyme inhibitor (see Wang *et al.*[13]).

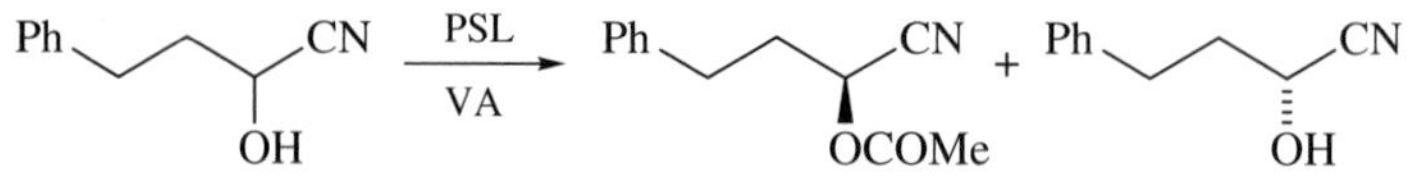

Scheme 3.4

Enzymic resolutions involve acceptance by the enzyme, which is a very finely honed chiral system, of one enantiomer of a racemic compound, but not the other. The selective acceptance arises because interactions between the enzyme and the enantiomers are diastereomeric. In its natural environment, the ability of an enzyme to discriminate between enantiomers is virtually absolute. In addition to their stereoselectivity, some enzymes can react at very high rates. Each round of catalysis by the enzyme carbonic anhydrase with its physiological substrate occurs in about 1.7 μs at room temperature, although for a small number of other enzymes, best exemplified by the more lethargic lysozyme, the corresponding figure is about a million times slower. Accordingly, the enzyme-catalysed hydrolysis of, say, one enantiomer of an ester proceeds at a finite rate and hydrolysis of the other not at all. Resolutions such as those of **39**, **42** and **45** therefore have a kinetic basis and are also known as kinetic resolutions.

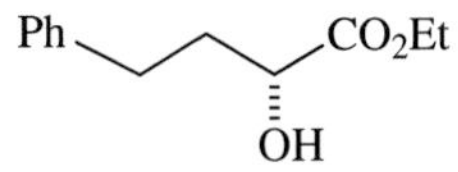

It should be evident that the maximum yield of a particular enantiomer normally available from a racemic mixture is 50%. However, in some enzymic catalysed kinetic resolutions it is possible to obtain >50% yield of one enantiomer from a racemate. For this to occur, it is necessary to have the desired chemical reaction, *e.g.* enzyme-catalysed stereoselective esterification, occurring at the same time as the enantiomers of the racemic starting compound are interconverting under equilibrium conditions. A successful example of this technique is provided by benzaldehyde cyanhydrin (2-hydroxy-2-phenylacetonitrile), whose *R* and *S* enantiomers, **49** and **50**, respectively, equilibrate in the presence of a basic anion-exchange resin (Scheme 3.5). In the presence of lipase, (*S*)-benzaldehyde cyanhydrin acetate **51** was formed in 95% yield and in 84% enantiomeric excess (see Inagaki *et al.*[14] and Ward[15]).

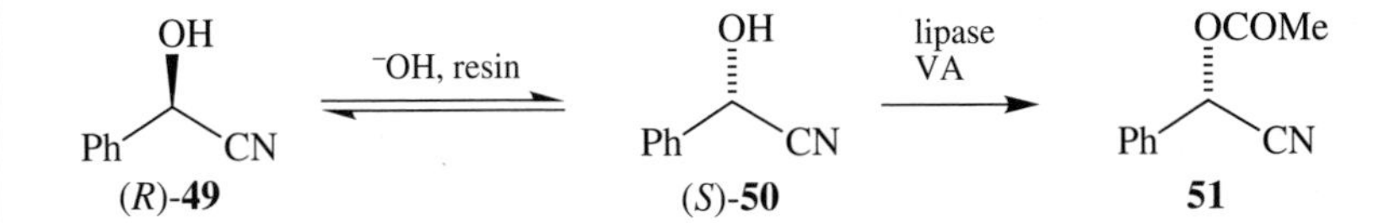

Yields of >50% of one enantiomer may be obtained from a racemic compound if, under the conditions of the enantioselective reaction, the enantiomers of the starting material are interconverting.

Scheme 3.5

As an alternative to resolution, one can start with an enantiomerically pure compound that occurs naturally, and the most common examples of these are amino acids, sugars and terpenes. This group of compounds is known collectively as the 'chiral pool'. The procedure then is to transform the compound of choice from the above group into the desired product by chemical synthesis, with care taken to avoid racemization at stereogenic centres.

3.9   Structure of Polypropene

Polymerization of propene gives polypropene (**52**), with creation of one stereogenic centre per monomer unit. Three possibilities exist for the structure of **52** and these are shown in the stereo drawings **53–55**. Compound **53** has a random distribution of configurations at carbons along the chain, and this type of polymer is known as **atactic**

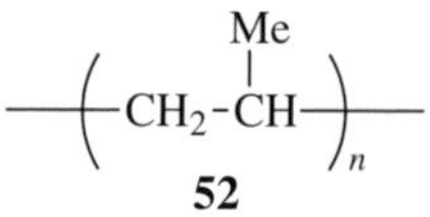

$$-\left(CH_2-\underset{\underset{Me}{|}}{CH}\right)_n-$$

52

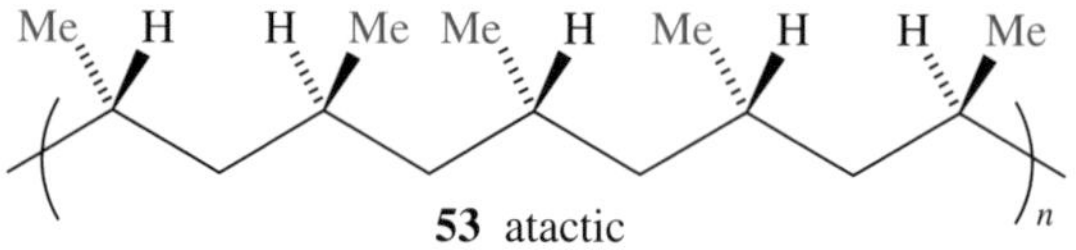

53 atactic

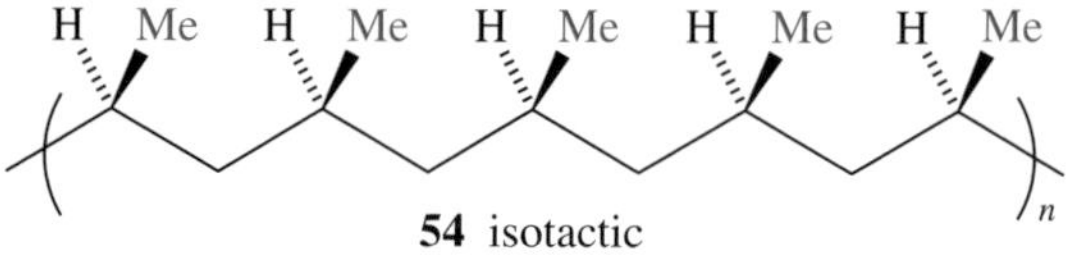

54 isotactic

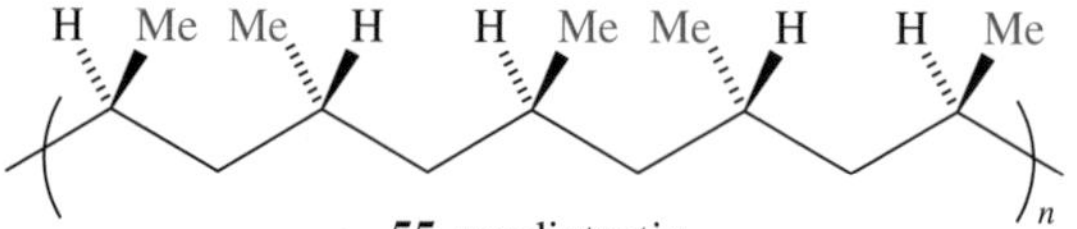

55 syndiotactic

In polymer **54**, all methyl groups are on the same side of the polymer backbone, which is drawn in the plane of the page; this type of polymer is known as isotactic. The methyl groups alternate regularly from one side of the chain to the other in polymer **55**, which is known as syndiotactic.

In general, atactic polymers tend to be soft, amorphous material, whereas the more regular structures of isotactic **54** and syndiotactic **55** polymers permit chains to lie closer together, and accordingly these polymers have a greater tendency to be crystalline. The terms atactic, isotactic and syndiotactic are quite general and are used for other similarly substituted polymers.

Worked Problems

Q1. Assign configuration to each of the stereogenic centres in **56**, **57** and **58**. Which of the compounds is/are *meso*?

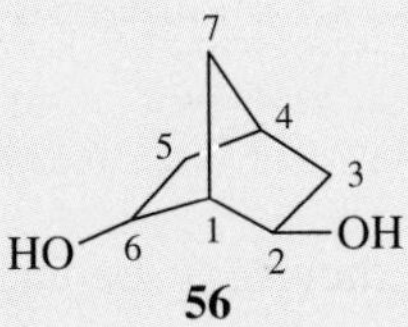

A. Compound **56** has configuration (2*R*,6*S*) and is (2*R*,6*S*)-bicyclo[2.2.1]heptane-2,6-diol, or with the *exo/endo* nomenclature, *exo,exo*-bicyclo[2.2.1]heptane-2,6-diol. This molecule clearly has a plane of symmetry that passes through C(1), C(7) and C(4), and inspection of the structure shows that C(2) and C(6) have opposite configuration; **56** is therefore *meso*. Compound **57** has configuration (2*S*,6*S*) and this molecule, lacking a plane of symmetry, is not *meso*. The acyclic diol **58** is (2*R*,4*S*)-pentane-2,4-diol, and like **56** is *meso*.

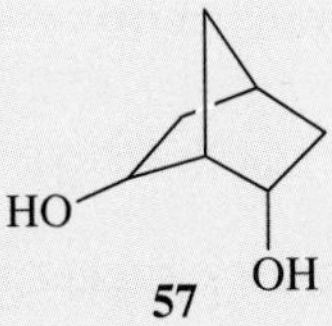

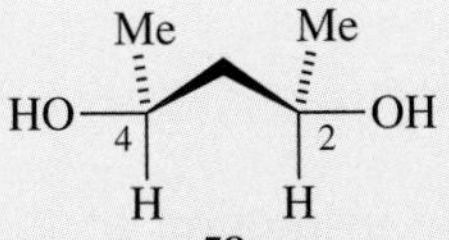

These answers demonstrate that compounds with appropriate stereochemistry, *i.e.* **56** and **58**, can be *meso* even though the stereogenic centres are not adjacent. One can see this more clearly in the rigid structure **56** than in the acyclic diol **58**. However, to demonstrate that these two molecules are conceptually related, proceed as follows: break the C(1)–C(7), C(3)–C(4) and C(4)–C(5) bonds in **56** to leave a diol, **58**, that derives from C(1), C(2), C(3), C(5) and C(6) of **56**, with, of course, hydrogens added to satisfy the tetravalency of carbon.

Q2. What is the stereochemical relationship between **59** and **60**? Assign *R/S* configurations at the stereogenic centres.

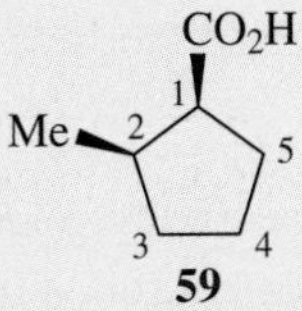

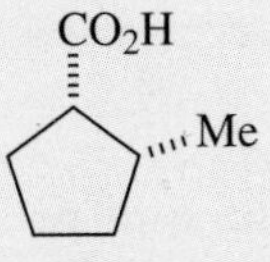

A. As with all problems of comparison, first look carefully at the two structures. Take **60** and rotate it by 180° around the axis that passes through C(1) and the mid-point of the C(3)–C(4) bond; then draw the result, which is **59**; the two structures are identical. The configuration is (1*S*,2*R*).

Q3. What is the stereochemical relationship between **61** and **62**? What are the *R/S* configurations of the stereogenic centres?

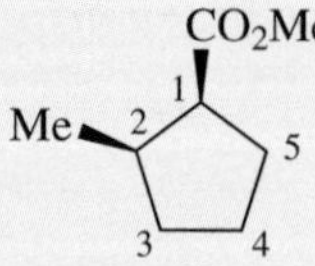

A. Inspection of the structures suggests that they may be enantiomers. Confirmation should be provided by molecular models, and also from the configurations of the stereogenic centres. These are (1*S*,2*R*) for **61** and (1*R*,2*S*) for **62**.

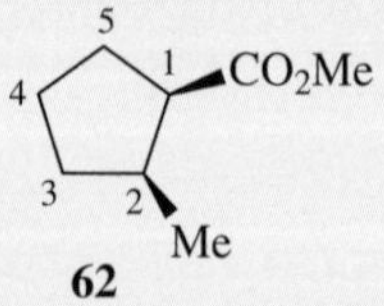

Q4. What is the stereochemical relationship between **63** and **64**? What are the configurations at C(1) and C(2)?

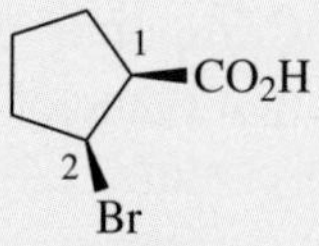

A. The configurations at C(1) in these molecules are different, whereas at C(2) they are the same. Bromo acid **63** has configuration (1*S*,2*S*) whereas **64** has configuration (1*R*,2*S*). These two compounds are related as diastereoisomers.

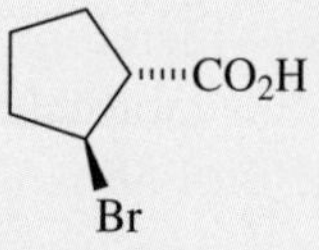

Q5. What is the stereochemical relationship between **65** and **66**? What are the configurations of the stereogenic centres, and the names of the molecules?

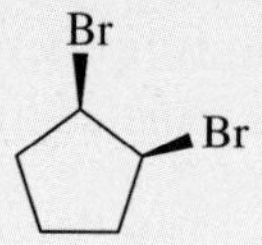

A. Compounds **65** and **66** are, respectively, (1*R*,2*S*)-dibromocyclopentane and (1*R*,3*S*)-dibromocyclopentane. The compounds are structural isomers; both are *meso*.

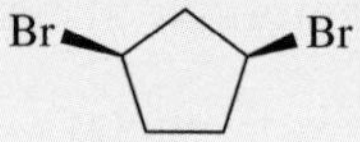

Q6. What is the stereochemical relationship between **67** and **68**?

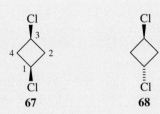

67 **68**

A. Neither **67** nor **68** is chiral: each has a plane of symmetry that passes through C(1) and C(3). These molecules are diastereoisomers (see Section 6.3; see also Section 6.5).

Q7. What is the stereochemical relationship between the pairs of compounds (i) **69** and **70** and (ii) **71** and **72**?

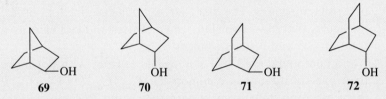

69 **70** **71** **72**

A. The pair of epimers **69** and **70** have opposite configuration at C(2). In **69** the hydroxyl group is on the same side as the shorter bridge and is termed *exo*; in **70** the hydroxyl is now on the same side as the larger bridge and is therefore *endo*.

In **71** and **72**, both unsubstituted bridges are equivalent and so there are no epimers. Instead, **71** and **72** are related as enantiomers; for these molecules this can be seen more easily (i) from molecular models and (ii) by a 120° anti-clockwise rotation around an axis that passes through the bridgehead carbons C(1) and C(4).

Q8. (–)-Carvone (**73**) occurs in spearmint and its enantiomer (+)-carvone (**74**) is found in caraway seed. To the human sense of smell, these two enantiomers have different odours. Where is the stereogenic centre in **73** and in **74**? What conclusions can be drawn about the human olfactory receptor site?

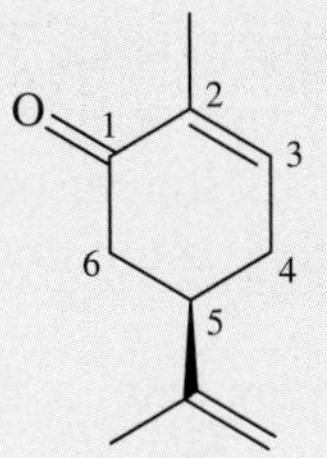

73 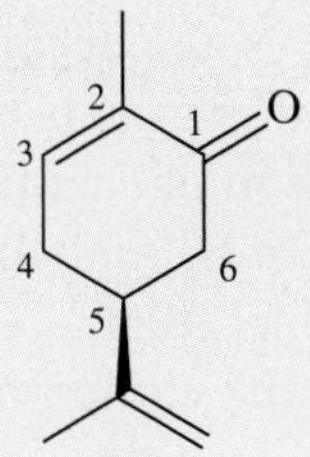 **74**

A. That different enantiomers **73** and **74** give rise to distinct odours means that the olfactory receptor site in humans is chiral. This has the consequence that the interactions of the enantiomers with the receptor site are diastereomeric and accordingly different odours are perceived. There are many pairs of enantiomers whose odour is identical, but differences between odours of enantiomers are most common in cyclic compounds (see Pybus and Sell[16]). The stereogenic centre in both **73** and **74** is at C(5).

Summary of Key Points

- Compounds with *n* stereogenic centres give rise to a maximum of 2*n* stereoisomers. The first four sections described, in order, molecules that have two stereogenic centres and can exist as a total of 4, 3 and 2 stereoisomers.
- For a compound with two stereogenic centres, with not more than one substituent in common, *i.e.* C_{abc}–C_{def} (or C_{abc}–C_{cde}), there are four stereoisomers with (using the *R/S* convention) configurations (i) *R,R*, (ii) *S,S*, (iii) *R,S* and (iv) *S,R*. Of these, (i) and (ii) are enantiomers, as are (iii) and (iv). Other combinations are diastereoisomers, that is stereoisomers that are not enantiomers.
- A compound with two stereogenic centres, but with the same substituents at each, *i.e.* C_{abc}–C_{abc}, is now restricted to three stereoisomers, *R,R*, *S,S* and *R,S*. The last of these is achiral, as is implied by stereogenic centres of opposite configuration, but each with the same substituents, and is called *meso*.
- The terms *erythro* and *threo*, and *syn* and *anti*, were described; the close relationship between *erythro* and *meso* is useful to help distinguish *erythro* from *threo*.
- Caged bicyclic compounds with stereogenic centres only at the bridgehead positions are restricted to two enantiomeric forms.
- Epimers are usually compounds with three or more stereogenic centres, but which differ in configuration at only one.
- Resolution is the separation of enantiomers from a racemic mixture and is usually achieved by one of three methods:
  1. Conversion of enantiomers into diastereoisomers, which are usually salts, and then exploiting solubility differences to effect separation.
  2. Differential elution of enantiomers from a chiral column.

3. Exploiting the ability of an enzyme to catalyse a reaction, usually either alkaline hydrolysis of an ester, or an esterification, of one enantiomer exclusively. If the enantiomers of the racemic starting material can be made to equilibrate while the enzyme-catalysed enantioselective reaction is taking place, yields of one enantiomer in excess of 50% can be obtained.

Problems

3.1. Draw all the stereoisomers of each of the molecules (a)–(e); assign configuration to stereogenic centres and say whether each stereoisomer is chiral: (a) 2,3-dibromobutane; (b) 2-bromo-3-chlorobutane; (c) the monomethyl ester of tartaric acid (2,3-dihydroxybutanedioic acid); (d) 2,3-difluoropentane; (e) 1,3-dichlorocyclopentane.

3.2. Draw a three-dimensional structure of a *meso* compound for each of the following formulae: (a) $C_2Br_2Cl_2F_2$; (b) C_8H_{18}.

3.3. With reasons, state whether each of the statements (a)–(c) is false or true: (a) a plane of symmetry in a molecule means that it is achiral; (b) all chiral molecules possess diastereoisomers; (c) all molecules with stereogenic centres are chiral.

3.4. Is it possible to resolve **75** with the aid of a chiral chromatography column?

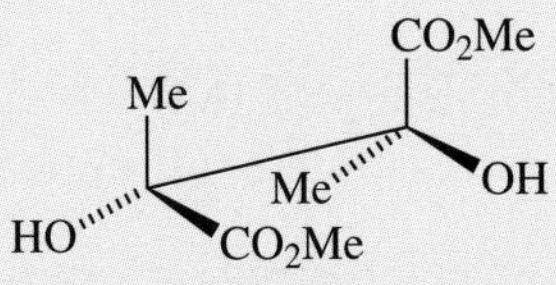

75

3.5. After a single enantiomer of each of **76** and **77** has been reduced with H_2/Pt, the product is optically inactive in one case. Which of **76** and **77** gives the optically inactive product, and why?

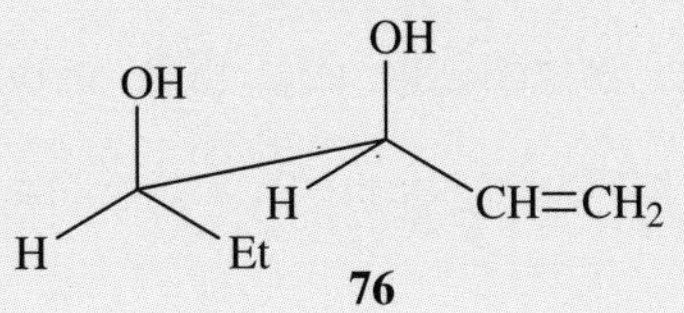

76 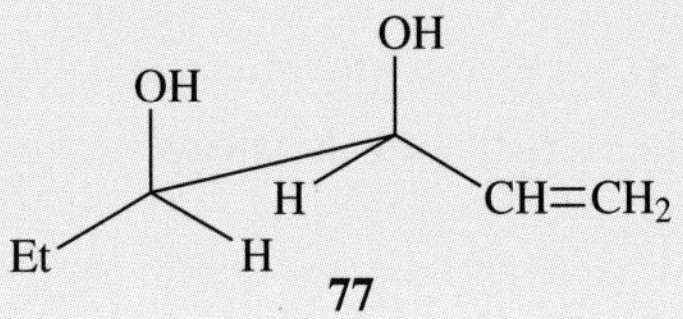   **77**

3.6. Give a stereo drawing or sawhorse projection of the *meso* stereoisomer of **78**.

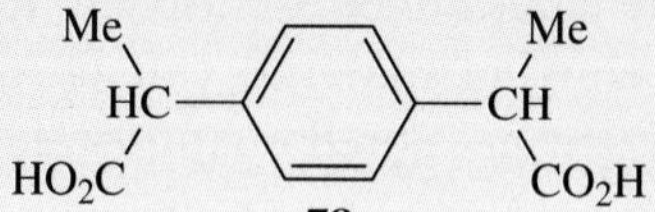

78

3.7. Identify the stereogenic centres in codeine (**79**) and in cholesterol (**80**).

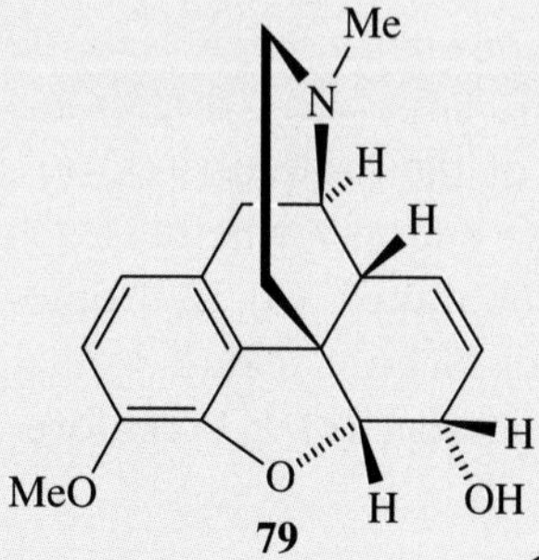

79

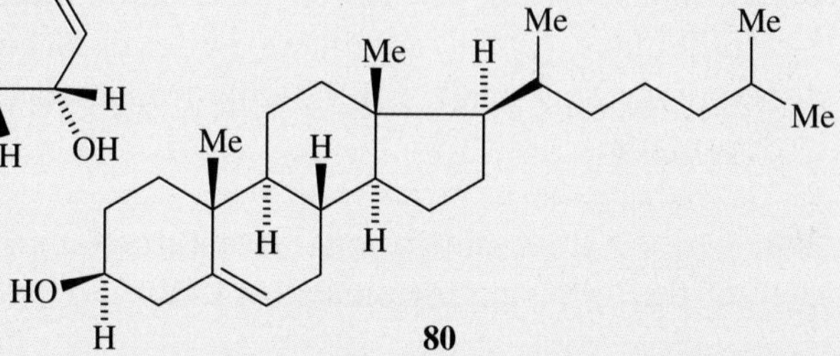

80

3.8. With reasons, state whether **81–83** are, or are not, chiral.

81

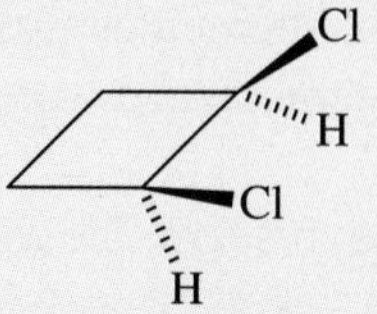

82

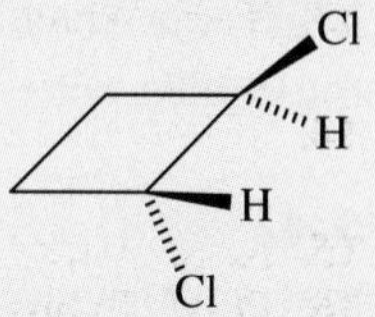

83

3.9. Explain how ricinoleic acid (**84**) has one stereogenic centre, yet can exist as four stereoisomers, all optically active.

$$Me(CH_2)_5 - \underset{\underset{OH}{|}}{CH} - CH_2 - CH = CH - (CH_2)_7CO_2H$$

84

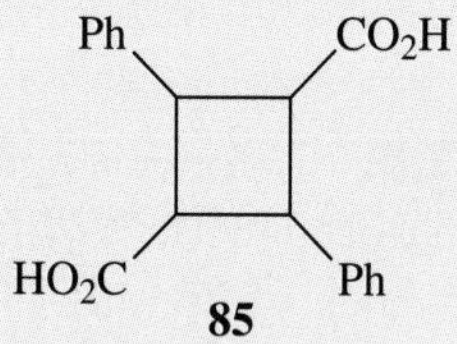

85

3.10. Draw the five stereoisomers of truxillic acid (**85**); why is none chiral (see Mislow[17])?

References

1. D. Seebach, R. E. Marti and T. Hintermann, *Helv. Chim. Acta*, 1996, **79**, 1710.
2. S. Masamune, S. A. Ali, D. L. Snitman and D. S. Garvey, *Angew. Chem. Int. Ed. Engl.*, 1980, **19**, 557.
3. D. Seebach and V. Prelog, *Angew. Chem. Int. Ed. Engl.*, 1982, **21**, 654.
4. G. N. Loudon, *Organic Chemistry*, 2nd edn., Benjamin Cummings, Menlo Park, California, 1988, p. 1137.
5. L. M. Harwood, C. J. Moody and J. M. Percy, *Experimental Organic Chemistry*, 2nd edn., Blackwell, Oxford, 1998, pp. 176, 182 (this text describes column chromatography)
6. W. L. F.Armarego and D. J. Perrin, *Purification of Laboratory Chemicals*, 4th edn., Butterworth-Heinemann, Oxford, 1996, p. 18 (this text describes column chromatography).
7. W. H. Pirkle, J. C. Pochapsky, G. S. Mahler, D. E. Corey, D. S. Reno and D. M. Alessi, *J. Org. Chem.*, 1986, **51**, 4991.
8. C. H. Lochmüller, in *Chiral Separations: Applications and Technology*, ed. S. Ahuja, American Chemical Society, Washington, 1997, p. 43.
9. L. Stryer, *Biochemistry*, 4th edn., Freeman, New York, 1995, p. 181.
10. C.-S. Chen, Y. Fujimoto, G. Girdankes and C. J. Sih, *J. Am. Chem. Soc.*, 1982, **104**, 7294.
11. S.-K. Kang, J.-H. Jeon, T. Yamaguchi, J.-S. Kim and B.-S. Ko, *Tetrahedron: Asymmetry*, 1995, **6**, 2139.
12. A. Chadha and M. Manohar, *Tetrahedron: Asymmetry*, 1995, **6**, 651.
13. Y.-F. Wang, S.-J. Chen, K. K.-C. Liu and C.-H. Wong, *Tetrahedron Lett.*, 1989, **30**, 1917.
14. M. Inagaki, J. Hiratake, T. Nishioka and J. Dola, *J. Am. Chem. Soc.*, 1991, **113**, 9360.
15. R. S. Ward, *Tetrahedron: Asymmetry*, 1995, **6**, 1475.
16. D. H. Pybus and C. S. Sell, *The Chemistry of Fragrances*, Royal Society of Chemistry, Cambridge, 1999, p. 68.
17. K. Mislow, *Introduction to Stereochemistry*, Benjamin, New York, 1966, p. 112.

4

Stereochemistry of Carbon–Carbon and Carbon–Nitrogen Double Bonds

---

Aims

By the end of this chapter you should be familiar with:

- The configuration, relative stability and nomenclature of alkenes, both acyclic and cyclic
- The configuration of the carbon–nitrogen double bond in, for example, oximes
- The structure and configuration of amides
- *Cis* hydroxylation of alkenes with OsO_4
- *Trans* hydroxylation of alkenes with peroxy acids to give oxiranes, which then undergo nucleophilic ring opening
- Acid-catalysed Markovnikov hydration of alkenes
- Anti-Markovnikov hydration of alkenes *via* intermediate organoboranes

4.1 Configuration and Relative Stability of Alkenes and Dienes

In 1,2-disubstituted alkenes the substituents can be on the same side of the double bond (*cis*) or on opposite sides (*trans*). Such compounds are geometric isomers.

Alkenes were considered in Chapter 1 in the context of hybridization. In ethene, all six atoms are co-planar. In general, one can state that alkenes have two adjacent sp^2 hybridized carbons, and these carbon atoms together with the four atoms to which they are attached are co-planar. Above and below the plane defined by these six atoms there exists a region of electron density, the π bond, which arises from sideways-on overlap of p orbitals on the adjacent alkene carbons.

In but-2-ene the two methyl groups can lie either on the same side or on opposite sides of the double bond. Compound **1** is *cis*-but-2-ene and compound **2** is *trans*-but-2-ene. Alkenes can be viewed either (a) from 'above', in which case the wedges and dashed wedges are not included, or (b) from almost sideways-on; in this case, wedges and dashed wedges are included. These indicate that substituents are directed toward or away from the viewer, respectively. Both representations are equally valid and are shown for *cis*-but-2-ene (**1**) in both ways. The second representation is more common when addition of reagents to the top (or bottom) face is described.

Compounds **1** and **2** can be regarded as non-interconvertible and are called **geometric isomers**; they also fulfil the definition of diastereoisomers, though they are only occasionally so called. The *cis* and *trans* nomenclature is unambiguous for disubstituted alkenes; however, the situation is not clear-cut for tri- and tetrasubstituted alkenes. With these latter compounds, problems arise in the definition of which substituent is *cis* or *trans* to which.

In order to circumvent this problem, Blackwood *et al.*[1] took part of the Cahn–Ingold–Prelog (CIP) convention, in particular the rules for assignment of priorities, and applied them to tri- and tetrasubstituted alkenes. In the case of the tetrasubstituted geometric isomers **3** and **4**, consider the substituents at C(2). In the CIP convention, Br ranks higher than C and so is given priority; at the other carbon, C(3), F ranks higher than C and it too is given priority. Now the two atoms, Br and F, which were assigned priority at the respective alkene carbons, are on opposite sides of the double bond. Accordingly, the geometric isomer **3** is given the symbol '*E*' from the German 'entgegen' for 'opposite'.

More generally, the CIP rules are adapted and adopted. The higher priority group at each sp^2 hybridized carbon is identified. If these are on the same side the double bond is *Z*, if on the opposite side it is *E*.

Using the same idea, in **4** the higher priority groups Br and F are now on the same side of the double bond. This geometric isomer is now given the symbol '*Z*', from the German 'zusammen' for 'together'. Compound **3** is named (*E*)-2-bromo-3-fluorobut-2-ene; similarly, **4** is (*Z*)-2-bromo-3-fluorobut-2-ene.

Of course, there is no objection to using this nomenclature to replace *cis* and *trans* in disubstituted alkenes, even where *cis* and *trans* is adequate. The disubstituted alkenes **5** and **6** are named with the systematic (*E/Z*) nomenclature.

5 (Z)-pent-2-ene   **6** (E)-pent-2-ene

Earlier in this section we mentioned that **1** and **2** were essentially non-interconvertible, and this is generally true for alkenes. However, it is worth keeping in mind that isomerization *can* occur in certain cases, and this is usually brought about by irradiation with light of a particular wavelength. An example is isomerization of *cis*- and *trans*-retinal, which is considered briefly in the Problems section.

If one considers the examples of **1** and **2**, or **5** and **6**, one might ask: which geometric isomer is more stable? Do the *cis* alkyl substituents get in each other's way at all, so that mutual non-bonded repulsion occurs with introduction of strain? The question has been addressed in the case of **1** and **2**, and answered by measurement of the enthalpy (heat) of hydrogenation of each alkene. This experiment is meaningful because both **1** and **2** give the same product, butane, after hydrogenation. Any difference in the enthalpies of hydrogenation is due to different enthalpy contents of the starting materials (enthalpies of hydrogenation do not, of course, measure free energies).

For (Z)-but-2-ene the enthalpy of hydrogenation is $\Delta H° = -119.6$ kJ mol^{-1}, and for (E)-but-2-ene the corresponding value is $\Delta H° = -115.5$ kJ mol^{-1}. Even though the difference between the two values is not large, it is significant. Since the (Z) geometric isomer **1** releases about 4 kJ mol^{-1} more heat on hydrogenation than does **2**, it follows that **1** is more strained because the methyl groups are closer together. Of course, with larger substituents, differences greater than 4 kJ mol^{-1} are encountered.

In molecules with two or more double bonds the configuration of each is determined independently, and the E/Z symbols are incorporated into the name as for **7** and **8**. In some dienes, one double bond has E/Z stereochemistry and the other does not because two groups on one carbon are the same. Such an example is **9**, which therefore has the name (Z)-3-chloropenta-1,3-diene.

The enthalpy of hydrogenation of the more congested, and hence strained, (Z)-but-2-ene is greater than for the E stereoisomer.

7 (2Z,4E)-hexa-2,4-dienoic acid   **8** (2E,4E)-hepta-2,4-diene   **9**

4.2   Cyclohexene

In cyclohexene (**10**) the enforced planarity of four carbons (and of course two attached hydrogens) means that the ring is not free to adopt the chair shape that is a characteristic of cyclohexane. The result is that cyclohexene exists as a half-chair, as shown in **10**, and this interconverts rapidly with **11**.

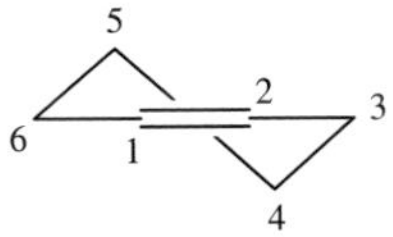

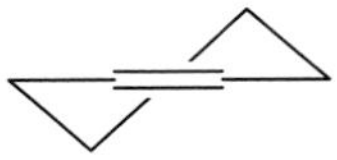

It is a good idea to make a molecular model of **10**. This model should indicate that whereas the hydrogens on C(4) and C(5) are essentially axial and equatorial, as in cyclohexane, those at C(3) and C(6) [carbons adjacent to the double bond] are recognizably different from cyclohexane, and so these allylic hydrogens are termed *pseudoaxial* and *pseudoequatorial*.

Cyclohexene has four coplanar carbon atoms and adopts a half-chair conformation. Allylic hydrogens adopt pseudo-axial and pseudo-equatorial positions, unlike hydrogens more remote from the double bond, which are axial and equatorial.

Cyclohexene is a *cis* alkene, and necessarily so, as the chain with four sp³ hybrid carbons is not long enough to link C(1) and C(2) with formation of a *trans* alkene. The smallest monocyclic *trans* alkene is *trans*-cyclooctene (**12**).

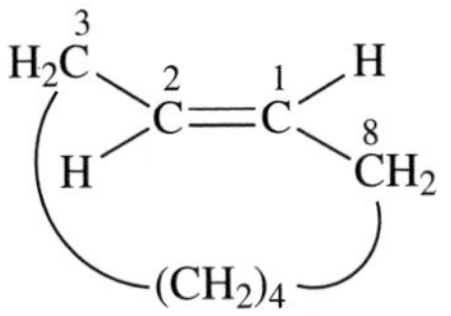

Make a model of this molecule with a flexible set of models; because the molecule is significantly strained, it is necessary to twist the double bond so that a Newman projection looks as in **13**. Then make a model of the mirror image of **13**; are they superimposable?

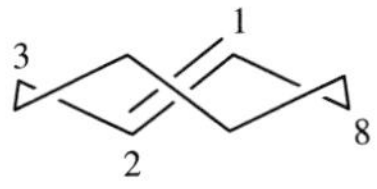

4.3 Carbon–Nitrogen Double Bonds

Carbon–nitrogen double bonds are present in imines **14**, and other carbonyl derivatives such as oximes **15**, semicarbazones **16** and dinitro-phenylhydrazones (DNPs) **17**. These carbon–nitrogen double bonds are formed by sideways overlap of p orbitals in the same manner as in alkenes. However, nitrogen is more electronegative than carbon, and imines are rather prone to hydrolysis. Compounds **14–17** are capable of existence as geometric isomers, as in the case of alkenes, and the lone pair on the imino nitrogen is generally taken to be approximately where a bond between sp² hybrid carbon and hydrogen is in an alkene.

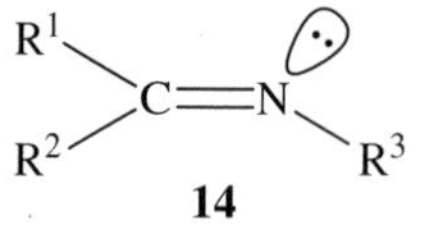

One can make an oxime, for example, from reaction between a ketone or aldehyde and hydroxylamine (NH_2OH), with liberation of water. It is rare to find that one has produced both geometric isomers of **15**. In the solid state, oximes nearly always exist as hydrogen-bonded dimers in which the hydroxyl hydrogen of one molecule forms a hydrogen bond with the nitrogen of its partner. The urea residue ($-NH-CO-NH_2$) in **16** establishes polar hydrogen-bonded networks; in the case of **17** the dinitrophenyl group is strongly polar. These features make for crystalline compounds, and so the three derivatives are used for characterization of ketones and aldehydes. The rates of interconversion of geometric isomers of, for example, oximes are very low, so that they are configurationally stable.

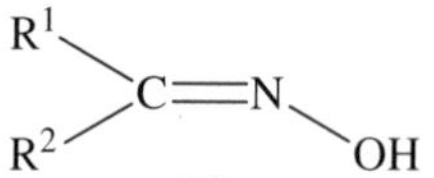

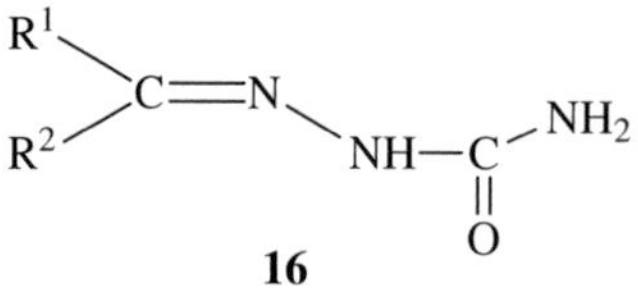

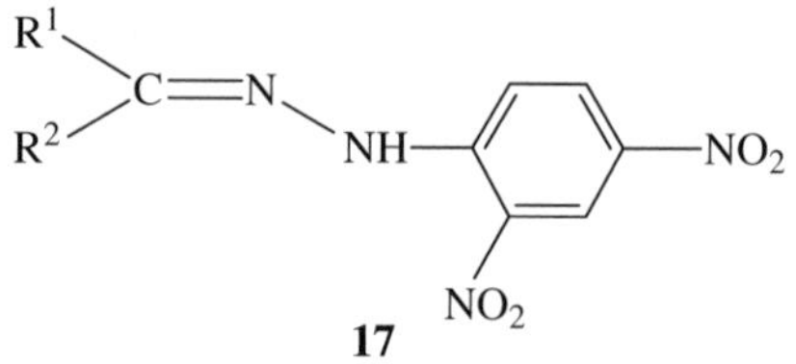

Carbon–nitrogen double bonds are configurationally stable, but this is only apparent when they are appropriately substituted.

4.4   Amides

Amides have six planar atoms, three (N, C, O) from the amide group and the three attached atoms.

In secondary amides, hydrogen and carbonyl oxygen are trans.

Amides have the general structure shown in **18**, in which R can be H, alkyl or aryl. The amide group is part of a six-atom planar entity, which contains, of course, the three atoms N–C=O together with the three atoms directly bonded to the N and to carbonyl carbon. In addition, the bond angles around N and C are 120°. In this section we consider mainly secondary amides, RHNCOR. The planarity and bond angles found in amides suggest that the hybridization of nitrogen in these compounds is sp². Now nitrogen is the strongest neutral conjugative electron donor, and the adjacent carbonyl group is correspondingly an excellent electron acceptor. Also, and importantly, the parallel alignment of the p orbital on N and the π bond of the carbonyl group, evidence for which comes from the six-atom plane of atoms mentioned above, means that extensive conjugation can occur. Accordingly, a secondary amide has the canonical structures shown in **19a** and the dipolar form **19b**. These can alternatively be represented by a single structure **20**, in which H and the two R groups are found to have the less congested *trans* relationship.

18

19a   **19b**

20

   As the carbonyl oxygen in an amide is more negatively charged than in, say, a ketone, it forms strong hydrogen bonds, but these must be either intermolecular or with a distant and suitable hydrogen in the same large molecule since intramolecular hydrogen bonding to H by the oxygen atom in the amide is normally precluded by the *trans* geometry. The importance of certain secondary amides in nature is well known, and this has its origins in the characteristics of structure **20**. Amides are the building blocks of polypeptides and proteins, and these adopt conformations that maximize intermolecular hydrogen-bonded interactions. One such conformation is a coil known as an α-helix, which is quite commonly found in peptide chains. A schematic representation of a right-handed α-helix is shown in Figure 4.1. Helical objects possess chirality, having essentially the same key features as a screw, and the right-handed helix

Secondary amides form strong intermolecular hydrogen bonds that in certain cases lead to the α-helix of polypeptides and proteins.

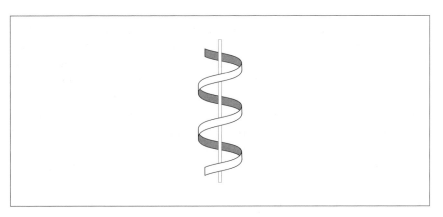

Figure 4.1 Representation of the right-handed α-helix of a polypeptide

is found with L-amino acids. Furthermore, the α-helix is closely associated with the structure of proteins. It is worth mentioning that the helix *itself* is inherently chiral, though chiral components with stereogenic centres are an integral part of its construction.

An interesting and relevant question is: how much double bond character is in the nitrogen–carbon bond in an amide? The carbon–nitrogen double bond is clearly sufficiently strong for a *trans* to *cis* interconversion not to occur at physiological temperatures, since this would disrupt the structures of polypeptides and proteins. The energetics of the *trans* to *cis* interconversion of amides have been investigated in a number of instances. We describe briefly the case of *N,N*-dimethylformamide (**21**, DMF; *N,N*-dimethylmethanamide), which can also be represented by the canonical form **22**. To the extent that **22** contributes to the actual structure, there is partial double bond character between N and carbonyl

Amides have partial double bond character between C and N; the rotational barrier around this bond is about 89 kJ mol⁻¹.

carbon. This has the effect of preventing rotation around the N–C bond, and as a consequence the two methyl groups are not in equivalent environments. This non-equivalence can be measured by the ¹H NMR spectrum of DMF, which at room temperature shows that the absorptions of the protons of the two methyl groups occur at different fields. However, as the temperature is raised sufficiently, rotation around the N–C(1) bond starts to occur and, after it has become sufficiently rapid, the magnetic shielding experienced by the two methyl groups becomes the same. At this point the discrete absorptions previously observed at room temperature are replaced by one unified signal. It is then possible to calculate the energy barrier to rotation, and in DMF this was found to be 89.1 kJ mol⁻¹. Values similar to this have been found for other amides.

This interconversion raises an interesting point. At room temperature, an amide represented by **20** (or **19b**), in the general case, can exist in two configurations, *trans* and *cis*, because of the partial C–N double bond. When the temperature is raised sufficiently, rotation starts to occur around the C–N bond, and conformational changes now take place. This is the only point in this text where such behaviour is found.

Here we note that the bond strength of a C–N double bond is *ca.* 615 kJ mol⁻¹ (about the same as a C–C double bond), and rather less than the *ca.* 748 kJ mol⁻¹ of a carbonyl group (see Sandorfy[2]).

4.5   *Cis* Hydroxylation of Alkenes

Hydroxylation of alkenes with osmium tetroxide, OsO_4, involves net addition of two hydroxyl groups to the same face of the alkene; this is *syn* addition. In the case of 1,2-dimethylcyclopentene (**23**), addition proceeds by way of a cyclic osmate ester **24**, whose formation is the result of a single-step addition to the alkene. Therefore, the *cis* nature of the methyl groups is maintained in **24** (see Scheme 4.1). The osmate ester is cleaved by $NaHSO_3$ (sodium hydrogensulfite), and this process involves breaking of both the single bonds between osmium and oxygen.

Alkenes are hydroxylated *syn* by either OsO_4 or MnO_4^-. In both cases the intermediates are cyclic esters, cleavage of which does not involve breaking the C–O bonds.

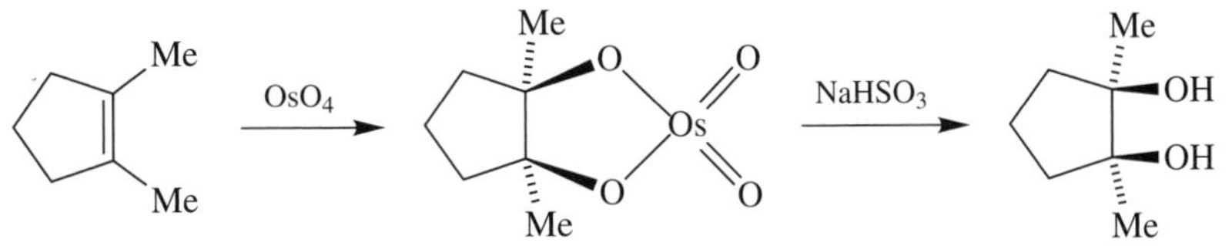

Scheme 4.1

The O–C bonds are not broken, and therefore the stereochemistry induced by formation of the osmate ester is maintained in the diol **25**. An analogous reaction occurs between **23** and permanganate [manganate(VII)] ion, MnO_4^-; now the corresponding intermediate is **26**, which breaks down in water to the *cis*-diol **25**, which is *meso*. Better yields are obtained with the more toxic reagent OsO_4; however, $KMnO_4$ is significantly less expensive.

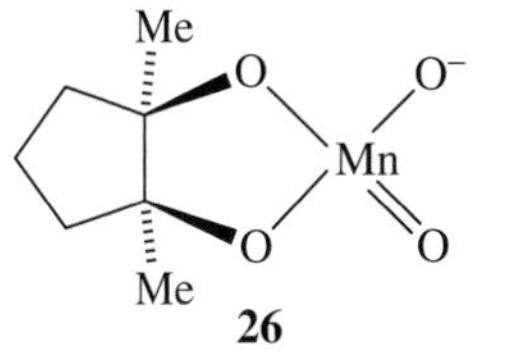

4.6   *Trans* Hydroxylation of Alkenes

Oxiranes (or epoxides) are conveniently formed by delivery of electrophilic oxygen to one of the faces of an alkene. This is achieved with a peroxy acid (peracid) in a single-step reaction (Scheme 4.2) that necessarily delivers oxygen *syn* to the double bond. Isotope labelling experiments have shown that the oxygen atom that is transferred to the alkene is the one that is further from the carbonyl group of the peroxy acid.

Trans hydroxylation of an alkene involves sequential conversion to an oxirane (epoxide), and its conjugate acid, followed by nucleophilic opening of the oxirane by water.

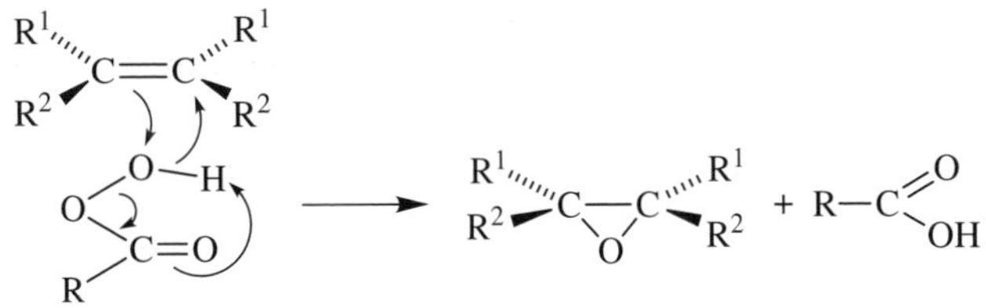

Scheme 4.2

In this reaction the C=C double bond acts as a nucleophile, and in most acyclic or monocyclic molecules oxygen is added with equal facility to the top or bottom face of the alkene. Oxiranes are the most reactive ethers, and are readily susceptible to nucleophilic attack, which results in ring opening. This is in contrast to diethyl ether (ethoxyethane), which is inert even in the presence of, for example, $LiAlH_4$. Likewise, 1,2-dimethylcyclohexene (**27**) reacts with peroxy acids to give *cis*-1,2-dimethylcyclohexene oxide (**28**).

Oxiranes, when protonated, are subject to nucleophilic attack by water to give a 1,2-diol as a ring-opened product. An example is shown by the reaction (Scheme 4.3) of the simplest oxirane, ethylene oxide (**29**), with water in the presence of acid to give initially the conjugate acid **30**. Nucleophilic attack by water, accompanied by ring opening, gives ethane-1,2-diol (**31**, ethylene glycol) (which is used as anti-freeze in cars).

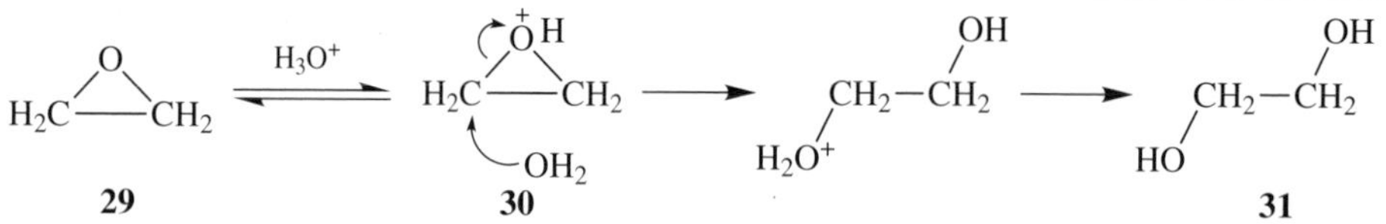

Scheme 4.3

Indeed, this is the method of choice for the industrial manufacture of **31**. The second step has wider relevance and accordingly is discussed briefly. In acidic solution, water is protonated to give H_3O^+, and the proton does not owe special allegiance to a particular water molecule, but instead migrates continually, from one water molecule to another (Scheme 4.4). Also present in the solution is **29**, and the proton will therefore spend a certain amount of time in **30**, having bonded to **29**. For the fraction of time that **30** exists in solution, it is susceptible to nucleophilic attack by water to produce **31**. This behaviour pattern is in operation for many acid-catalysed reactions of other oxygen-containing substrates.

Ring-opening of **32** proceeds along similar lines (see Scheme 4.5), *via*

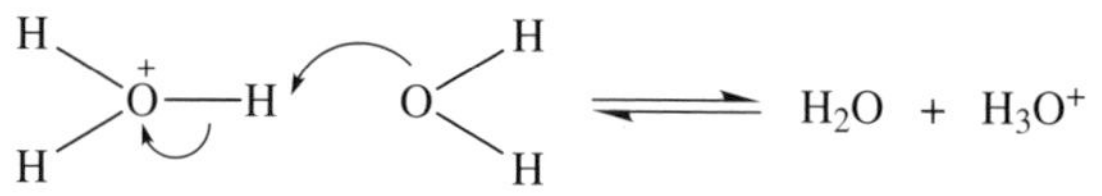

Scheme 4.4

a protonated intermediate which is subject to nucleophilic attack by water (S_N2) with inversion of configuration at C(1), with the result that the vicinal hydroxyl groups are *trans* in the diol **33**. This inversion of configuration is responsible for the difference in relative configuration of the diol function in **33** compared to those of the diols produced from alkenes by either OsO_4 or $KMnO_4$.

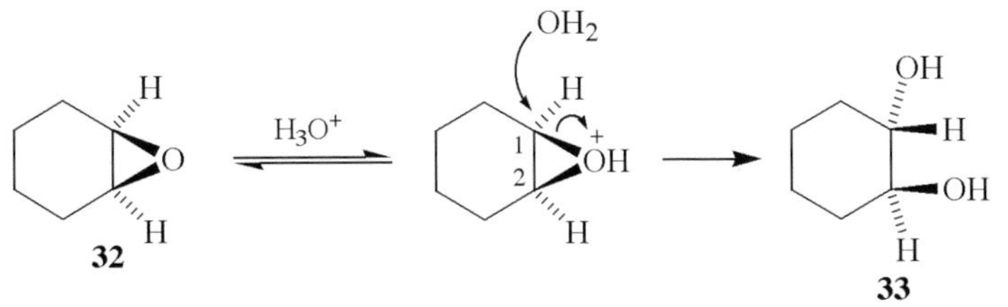

Scheme 4.5

In the previous reaction scheme an oxirane was converted into its conjugate acid by protonation, and in so doing a better leaving group was created. However, oxiranes, especially those with a primary carbon, will react with aqueous sodium hydroxide preferentially at the less-hindered primary carbon to give diols, as shown in the reaction of **34** to produce **35** (Scheme 4.6). In general, acidic conditions are preferred for the ring opening of oxiranes to produce *anti*-1,2-diols from alkenes.

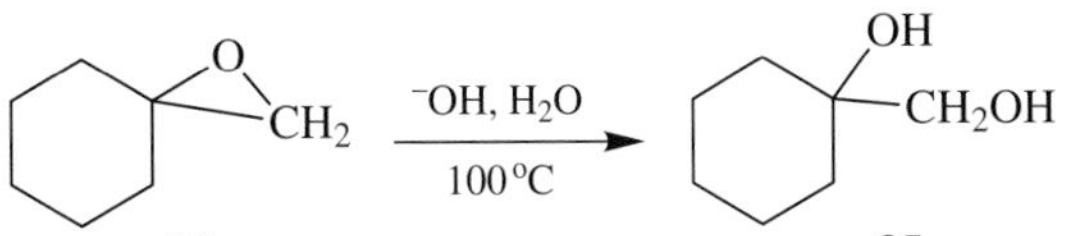

Scheme 4.6

4.7 Addition of Bromine to Alkenes

Alkenes react with bromine to give the products of 1,2-addition. The reaction is classified as an electrophilic addition of bromine, and the two bromine atoms in the product, 1,2-dibromoalkane, are mutually *trans*. Therefore from the addition of bromine to *trans*-but-2-ene (**2**) the product is *meso*-2,3-dibromobutane (**37**). This result is explained as follows: the initial step is nucleophilic attack by the double bond of the alkene on one bromine, with displacement of the other as bromide ion. The organic intermediate is a bromonium ion **36**, whose formation is rationalized in Scheme 4.7. The bromide that was expelled in formation of **36** now becomes a nucleophile and attacks **36** with equal probability at C(2) or C(3). In each case the reaction occurs with inversion of configuration to produce **37**.

Trans 1,2-bromination of an alkene proceeds *via* a bromonium ion; this cyclic intermediate is converted into the product by nucleophilic ring opening, this time by bromide.

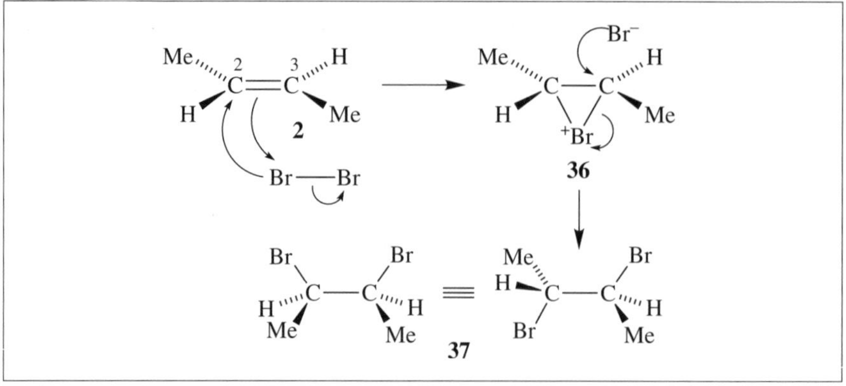

Scheme 4.7

With molecular models, verify that an analogous nucleophilic attack at C(2) of **36** gives the same *meso* compound. Confirm also that the configuration of **37** is (2*R*,3*S*). Show that cyclopentene (**38**) reacts similarly *via* a bromonium ion intermediate to give *trans*-1,2-dibromocyclopentane (**39**).

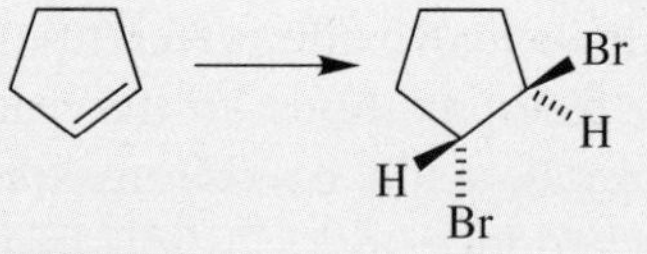

In Scheme 4.8 we consider the related reaction of (*Z*)-but-2-ene (**1**) with bromine. The intermediate bromonium ion **40** is similarly converted into **41**, which has the configuration (2*R*,3*R*), and **42** which has the configuration (2*S*,3*S*), and since equal amounts of these enantiomers are formed, the product is racemic. The same conclusion is obtained if the bromonium ion is formed on the top side of **1**, as drawn. That **41** and **42** are enantiomers can be seen from Newman projections, looking in the C(3)–C(2) direction. Noteworthy features of the electrophilic addition of

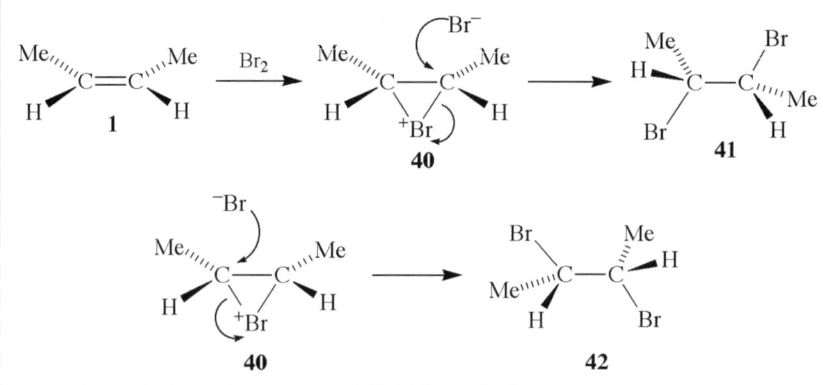

Scheme 4.8

bromine to alkenes are: (i) the π bond acts as a nucleophile toward bromine in the first step to give a bromonium ion; (ii) in the product-forming step, nucleophilic attack of bromide ion at C(3) [or C(2)] proceeds with inversion of configuration.

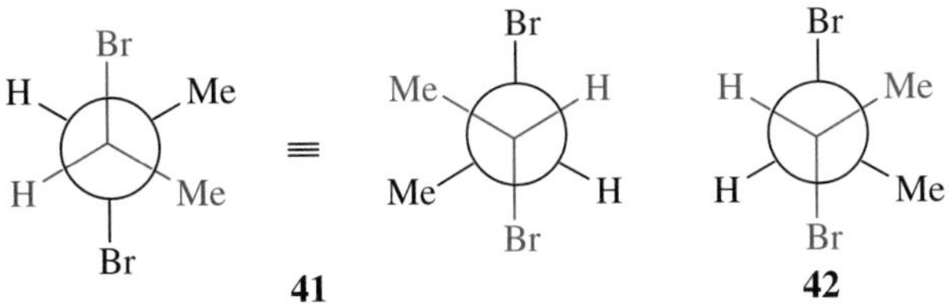

41 **42**

Consistent with the mechanism proposed, Strating et al.[3] isolated a bromonium ion from addition of bromine to the hindered alkene **43**; in fact the hindrance was such that the second stage, which would normally give the vicinal dibromoalkane product, did not occur. For an X-ray crystallographic structure determination of this bromonium ion, see Slebocka-Tilk et al.[4]

In the bromination of an alkene, one might ask: can bromine approach the alkene sideways on, rather than from either the top or bottom? Consider firstly the electron density of the double bond; this is directed well away from, and perpendicular to, the plane of the molecule, as described in Chapter 1.

From the congested diene **43**, a stable bromonium ion has been isolated, and its structure determined.

Secondly, make a model of the hindered alkene **43**; this shows that it is sterically impossible for bromine to bond with the alkene carbons other than by a top or bottom approach. To construct a molecular model of **43**, make two models of bicyclo[3.3.1]nonane, with both cyclohexane rings in a chair conformation. Identify carbons C(3) and C(7) in each of these structures; then take a 'molecule' of ethene and use C(1) of ethene to bond to C(3) and C(7) of one bicyclic structure, and C(2) of the alkene to bond to C(3) and C(7) of the other structure. The result is the hindered alkene **43**. Examination of this model, together with an understanding of the orientation of the double bond, indicates that steric factors permit isolation of the bromonium ion.

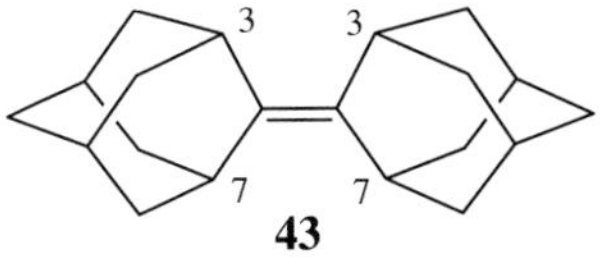

43

4.8   Hydration of Alkenes

4.8.1   Markovnikov Orientation

Markovnikov hydration, brought about by aqueous acid, involves a carbocation intermediate. The product has hydrogen attached to the less alkylated carbon.

In conventional hydration of alkenes by water in the presence of a strong acid, there is little direct stereochemical consequence, but we mention it as a useful contrast to the content of the next section. A proton initially adds to the alkene, and in the case of 2-methylpropene (**44**) gives the intermediate carbocation **45** (Scheme 4.9), which is then captured by the nucleophile, water, and yields the product 2-methylpropan-2-ol (**46**).

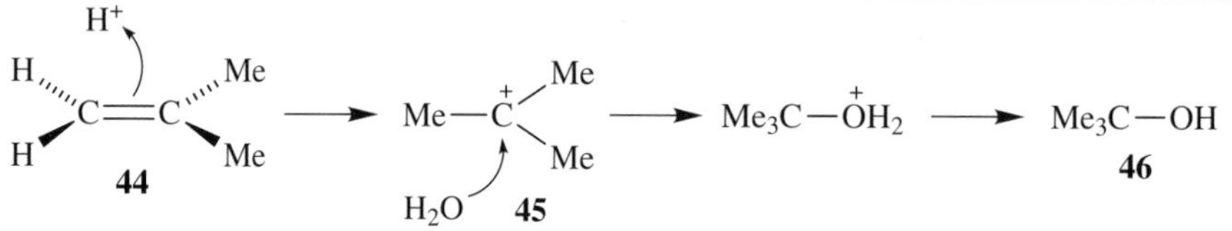

Scheme 4.9

A key point is that in the first step the proton is bonded to the less alkylated carbon. This is because the carbocation **45** is tertiary and stable. The alternative possibility is to produce a primary carbocation $Me_2CHCH_2^+$ that is significantly less stable. The order of relative stabilities of carbocations is tertiary > secondary > primary, and it is this relative sequence that dictates the site of proton attachment. Accordingly, one can state **Markovnikov's rule**, proposed in 1869: in the addition of H–X to a double bond, H becomes bonded to the carbon with fewer alkyl groups and X becomes attached to the more alkylated carbon.

A further example concerns reaction of but-1-ene (**47**) with aqueous acid. From this example (Scheme 4.10) the more stable carbocation is **48**, which reacts with water to give butan-2-ol (**49**). The carbocation **48** is planar and water can approach from either side with equal ease and so the product is racemic.

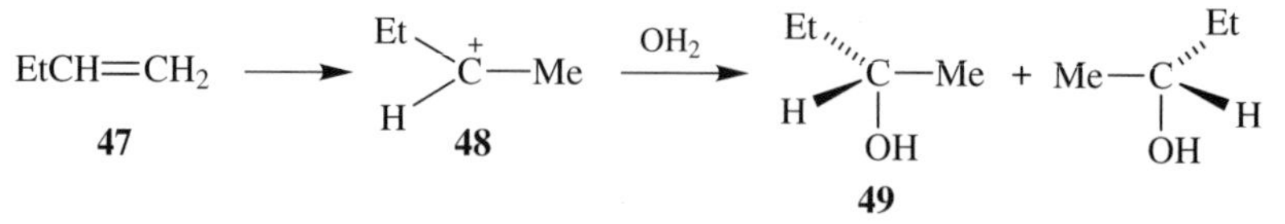

Scheme 4.10

4.8.2   Anti-Markovnikov Hydration

Another possibility is addition of the H and OH of water to an alkene so that H bonds to the carbon with more alkyl groups, and OH to the less alkylated carbon. Clearly intermediate carbocations are to be avoided. Such **anti-Markovnikov addition** has been achieved by a two-stage reaction: (i) the first stage is addition of the elements of borane, BH_3, across the double bond to give an organoborane; (ii) this organoborane is then transformed into an alcohol.

Hydroboration is usually carried out in the ether tetrahydrofuran, in which borane exists as a complex **50**, from which BH_3 is added to an alkene, *e.g.* 2-methylpropene (**44**) in Scheme 4.11. Addition takes place at a face of the alkene by means of a four-centre transition state, as shown in **51**. The partial bonds in **51** represent progressive formation of bonds between C and H, and between C and B, together with simultaneous weakening of the π bond and the B–H bond. In Scheme 4.11 the reaction of borane **52** is detailed; this borane has two remaining B–H bonds, and a similar reaction of these two bonds with two further molecules of alkene results in exhaustive alkylation, with formation of the trialkylborane **53**. The nature of the transition state **51** implies that H and B are delivered *syn* (to the same face), and simultaneously, to the double bond.

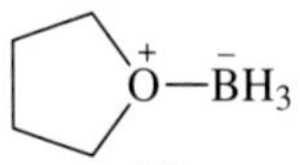

Anti-Markovnikov hydration is achieved by means of hydroboration with BH_3. Hydrogen adds to the more alkylated carbon, and BH_2 to the less alkylated, from the less hindered side in a reaction that proceeds *via* a four-centred transition state.

50

All three hydrogens of BH_3 are involved in the addition, with resultant formation of a trialkylborane. In conversion of the resultant organoborane to alcohol (by ^-OH, H_2O_2), oxygen replaces boron with retention of configuration at carbon.

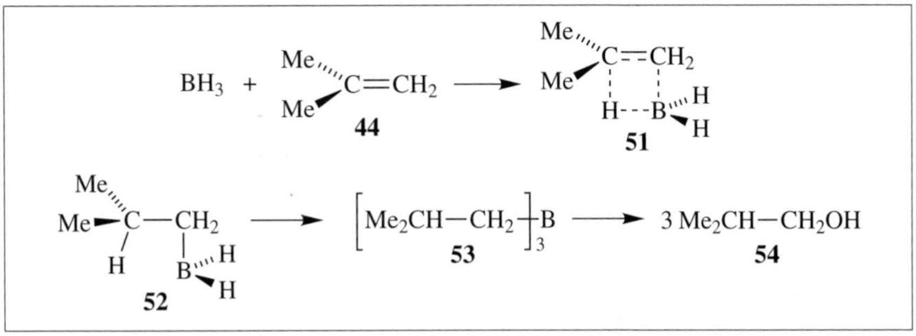

Scheme 4.11

The trialkylborane **53** is converted to alcohol **54** by reaction with an aqueous solution containing ^-OH and H_2O_2 (Scheme 4.12). Inspection of the structures of the alkene **44** and the alcohol **54** shows that addition of water has taken place in an anti-Markovnikov sense; in particular, hydrogen has now been added to the more alkylated carbon. One important feature of the hydroboration is transformation of **53** into **54**, which occurs with retention of configuration.

Scheme 4.12

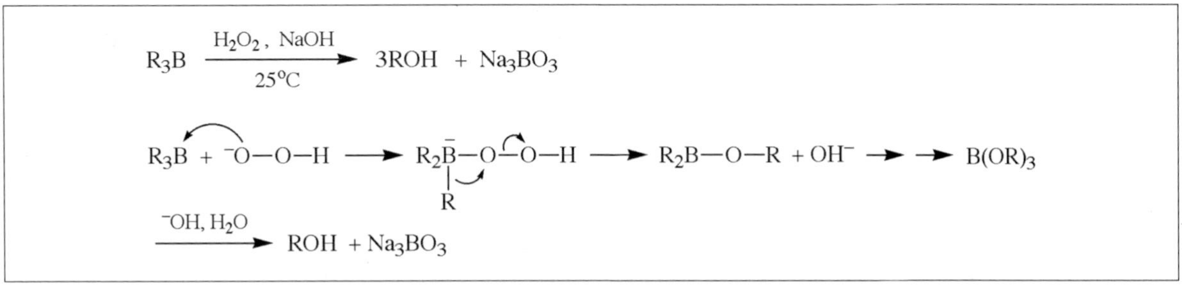

The examples given below illustrate the synthetic utility of the hydroboration reaction for anti-Markovnikov hydration of the double bonds in compounds **55–58**, with formation of alcohols **59–62**, and with predominant net delivery of the elements of water to the less-hindered alkene face, where relevant. Alcohols **61** and **62** are *threo* and *erythro*, respectively. *Cis* hydration of the double bonds of cyclic alkenes is described in the papers of Brown *et al.* and Alldred *et al.* and the text of Pelter *et al.* (see Further Reading).

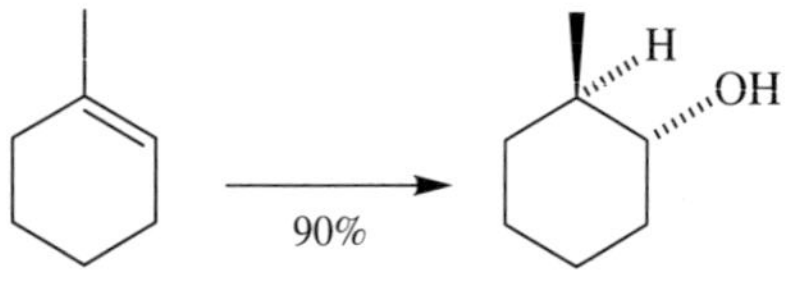

55 **59**

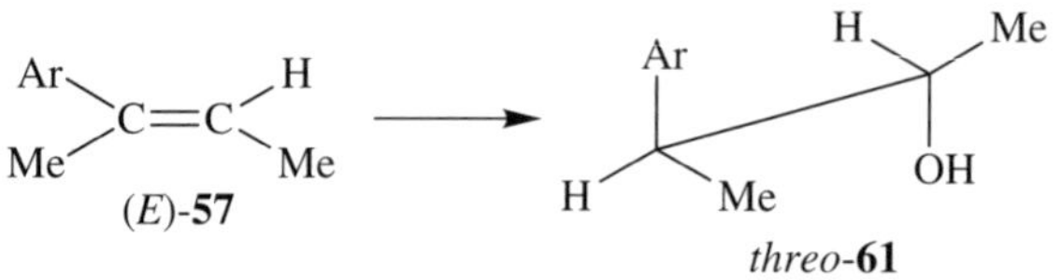

(*E*)-**57** *threo*-**61**

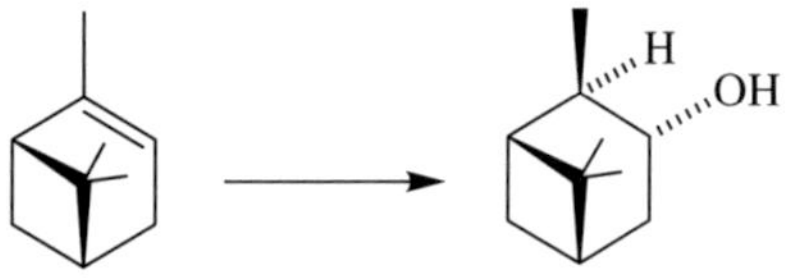

56 **60**

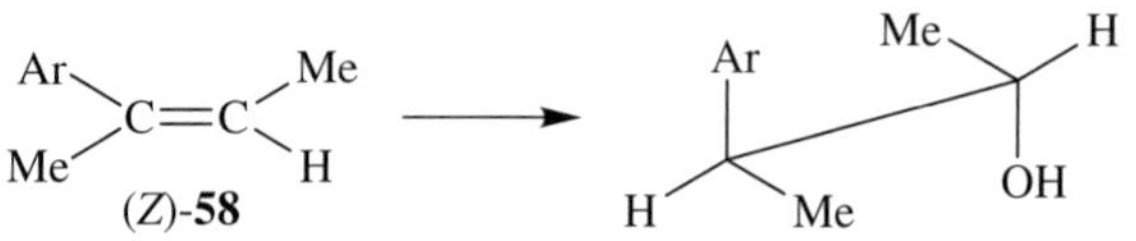

(*Z*)-**58** *erythro*-**62**

Worked Problems

Q1. *meso*-3,4-Dihydroxyhexane can be made from two alkenes that are geometric isomers. Outline two methods that make this possible.

A. *Vicinal* dihydroxy alkanes are available from alkenes by two routes: (i) peroxidation, followed by protonation and ring opening; (ii) formation of an osmate ester, followed by ring opening by treatment with $NaHSO_3$. The pathway in (i), which resembles that for bromination of an alkene, makes the (*E*)-alkene **63** a likely starting material. The mechanism, with inversion of configuration in the product-forming step, is given in Scheme 4.13.

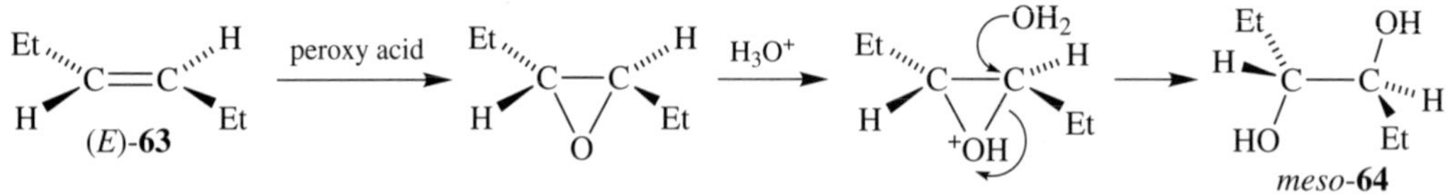

(*E*)-**63** *meso*-**64**

Scheme 4.13

The second pathway, outlined in (ii), does not involve inversion of configuration. Accordingly, the (*Z*)-alkene **65** is the one that gives the *meso* diol, as shown in Scheme 4.14.

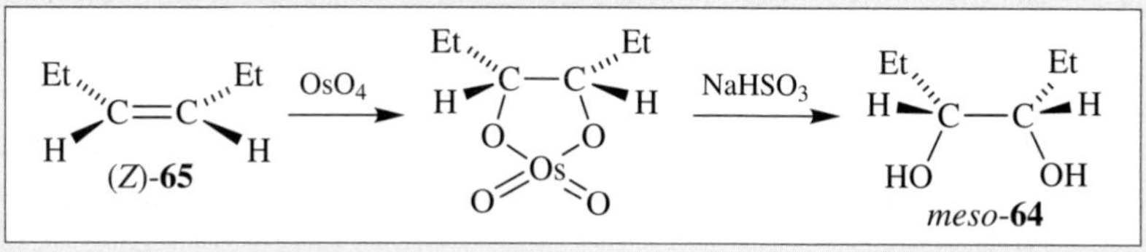

Scheme 4.14

Q2. What is the stereochemical nature of the product from the (*E*)-alkene **63** (Scheme 4.15) after sequential treatment with OsO_4 and $NaHSO_3$?

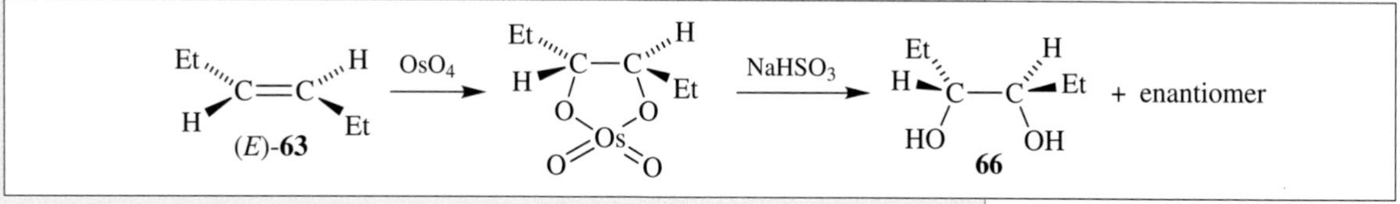

Scheme 4.15

A. The mechanism in Scheme 4.15 is the same as that in Scheme 4.14; the only difference is that the alkenes in these two schemes are geometric isomers. If one forms the osmate ester from the top face of **63**, then the enantiomer of **66** is obtained. The 'top' and 'bottom' faces of **63** are osmylated with equal probability; this means that the product is racemic.

Q3. When bromine is added to an alkene in solution in CCl_4, a dibromide is produced. Explain why, when bromine is added to the alkene **1** in aqueous solution, the product is a halohydrin, *i.e.* HO and Br have been added to adjacent atoms. What is the nature and stereochemistry of the product?

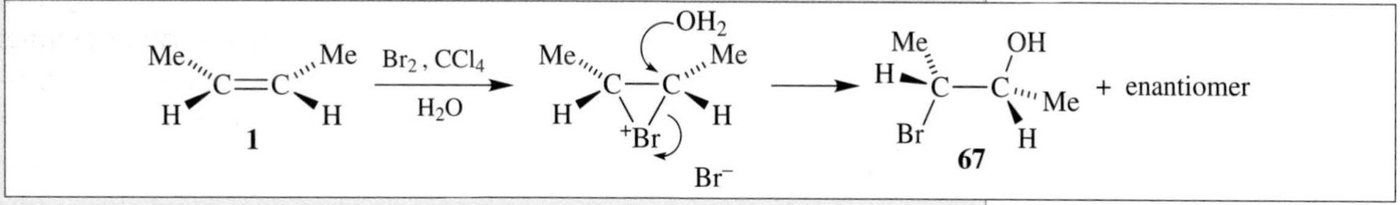

Scheme 4.16

A. The mechanism of the reaction is shown in Scheme 4.16; the first step (attack shown only on the lower face) gives a bromonium ion, as in bromination (Scheme 4.7). Present in the solution as nucleophiles are: (a) an equimolar amount of Br⁻, and (b) a stoichiometric excess of water (bulk water has a concentration of about 55 mol dm⁻³). Water is the effective nucleophile, and gives a product in which configuration has been inverted at the carbon attacked; bromide ion behaves similarly as shown in Scheme 4.7. The product **67** is formed in equal amounts with its enantiomer, and so it is racemic.

Q4. In the presence of a catalyst such as Pt, addition of hydrogen to an alkene brings about reduction to an alkane. Hydrogen is adsorbed on to the surface of the catalyst, and is then added *syn* to the alkene in a reaction that proceeds *via* a four-centre transition state **68** (*cf.* hydroboration, Section 4.8.2). What are the products of reduction of compounds **69** and **70** with H_2/Pt and are the products racemic or *meso*? Also, what is the product from reduction of **1** with D_2/Pt, and is this compound racemic or *meso*? What is the configuration of C(2) in the product?

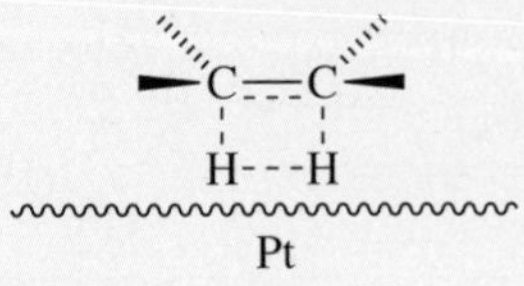

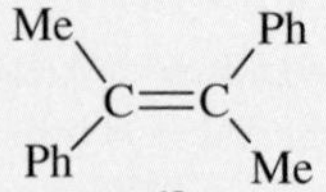

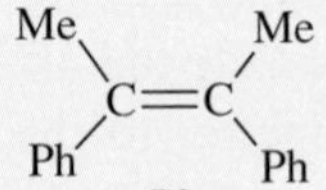

A. With a transition state as in **68**, the product from **69** is **71** and its enantiomer, in equal amounts. The product is therefore racemic. Compound **72**, the product of hydrogenation of **70**, is *meso*, as is the dideuteriated alkane **73**. The configuration of C(2) in **73** is (*R*).

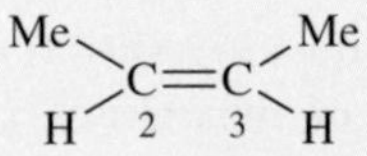

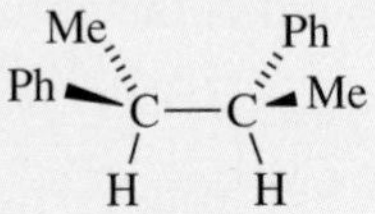

Summary of Key Points

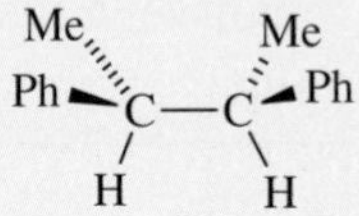

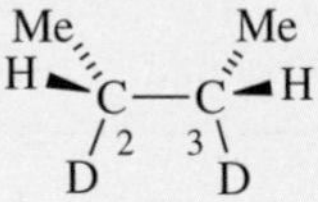

- Since there is no rotation around a carbon–carbon double bond C(1)=C(2), two isomers can exist when the double bond carries different substituents at C(1) and C(2).
- If the two substituents are on the same side of the double bond the isomer is termed *cis*, whereas if they are on the opposite side the isomer is *trans*.
- *Cis* and *trans* isomers are geometric isomers (and also can justifiably be called diastereoisomers).

- The priority sequencing protocol of the Cahn–Ingold–Prelog convention is used for di-, tri- and tetrasubstituted alkenes. The higher priority substituent at C(1) is identified, as is that at C(2). If the two higher priority substituents are on opposite sides of the double bond, the alkene configuration is (*E*); if they are on the same side, the configuration is (*Z*). The convention is also applied to multiple double bonds.

- (*E*)-But-2-ene is more stable than its (*Z*) geometric isomer by about 4 kJ mol^{-1}.

- Cyclohexene has four planar carbon atoms, a property that dictates that the molecule adopts a half-chair conformation. The allylic hydrogens are pseudo-equatorial and pseudo-axial, and the remaining four are essentially equatorial and axial.

- Carbon–nitrogen double bonds exist in carbonyl derivatives, *e.g.* oximes. When nitrogen carries an alkyl group, the imine so formed is prone to hydrolysis.

- In amides, six atoms, those of the N–C=O group and the three attached atoms, are coplanar with bond angles of 120°.

- When one of the groups on nitrogen is hydrogen, the hydrogen is *trans* to the carbonyl group and intramolecular hydrogen bonding to the carbonyl group (which is more polar than in a ketone) is normally excluded. The implication is that such hydrogen bonding is either intermolecular or occurs between well-separated centres in the same molecule, and in physiological systems this leads to the α-helix observed in polypeptides and proteins.

- The bond between nitrogen and carbonyl carbon in amides is configurationally stable at room temperature, but being only a partial double bond, rotation around the σ bond occurs at elevated temperatures, and the rotation barrier can be measured by means of ^{1}H NMR spectroscopy.

- Dihydroxylation of alkenes, the addition of two OH groups, can occur *cis* when either OsO_4 or MnO_4^- is used; cyclic esters are implicated.

- Alkenes are *trans* hydroxylated by a two-step sequence in which an intermediate oxirane is opened with inversion of configuration, most commonly by aqueous acid.

- Electrophilic addition of bromine to alkenes similarly involves a 1,2-*trans* addition.

- Alkenes are hydrated by aqueous mineral acid in a process that involves a carbocation. The less alkylated carbon accepts the proton, and the more alkylated receives the OH; this is one formulation of Markovnikov's rule.

- Anti-Markovnikov hydration of alkenes involves hydroboration, a reaction in which BH_3 adds to the alkene from the less
  hindered side in a four-centre reaction . Hydrogen adds to the
  more alkylated carbon, and boron to the less alkylated carbon.
  The net hydration is completed by replacement of boron with
  hydroxyl in a step that is characterized by retention of configuration at carbon.

Problems

4.1. By means of the *E/Z* convention, assign configurations to the
unsaturated compounds **74–80**.

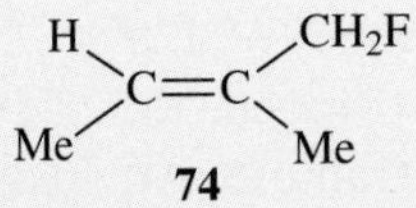

74

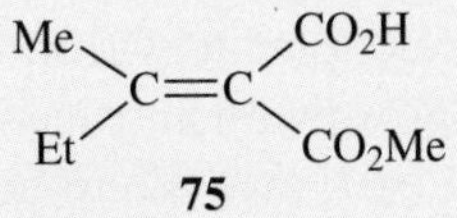

75

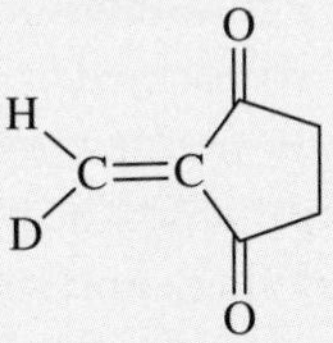

76

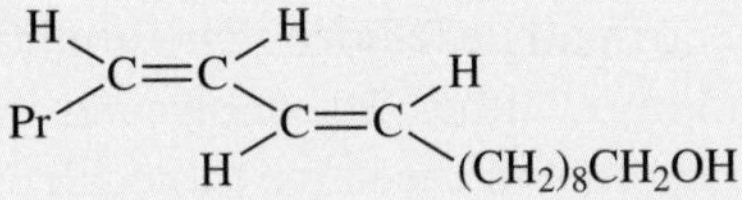

77 (sex attractant of silkworm moth)

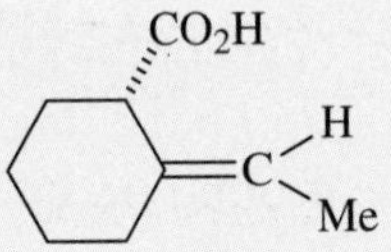

78

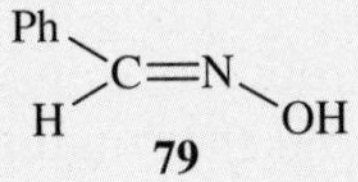

79

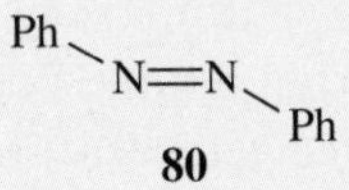

80

4.2. Nerol (**81**) is a naturally occurring compound called a
monoterpene (monoterpenes all contain 10 carbon atoms), present
in oil of bergamot. Assign stereochemistry to its double bonds.

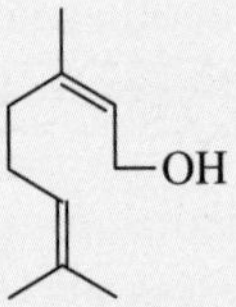

81

4.3. In the mechanism of vision, *cis*-retinal (**82**) is isomerized under the influence of light into *trans*-retinal (**83**). Give the *E/Z* stereochemistry of the double bond that undergoes isomerization in (a) **82**, before it is stimulated by light, and in (b) **83**, after stimulation and isomerization.

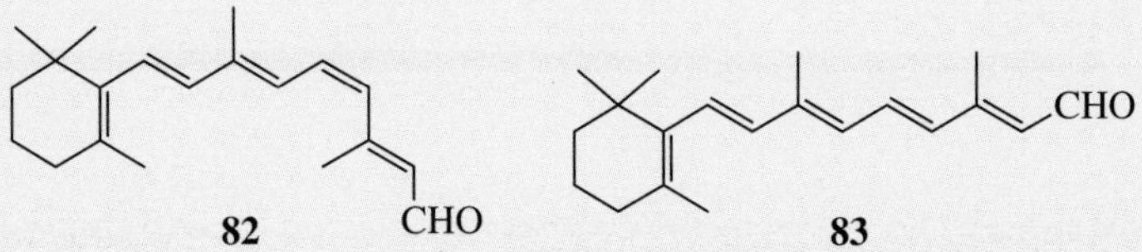

4.4. Draw stereochemical diagrams for: (i) (2*E*,4*Z*)-hexa-2,4-diene; (ii) (*E*)-2-bromopent-2-ene; (iii) (*Z*)-3-bromohexa-1,3-diene; (iv) 1,1-dichloropropene.

4.5. What are the products of reaction of **84** with: (i) bromine in CCl_4; (ii) OsO_4, then $NaHSO_3$; (iii) 3-chloroperbenzoic acid, then aqueous acid; (iv) H_2/Pt.

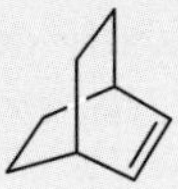

84

References

1. J. E. Blackwood, C. L. Gladys, K. L. Loening, A. E. Petraca and J. E. Rush, *J. Am. Chem. Soc.*, 1968, **90**, 509.
2. C. Sandorfy, in *The Chemistry of the Carbon–Nitrogen Double Bond*, ed. S. Patai, Wiley, New York, 1970, p. 1.
3. J. Strating, J. H. Wieringa and H. Wynberg, *Chem. Commun.*, 1969, 967.
4. H. Slebocka-Tilk, R. G. Ball and R. S. Brown give an X-ray structure determination of the bromonium ion derived from **43** in *J. Am. Chem. Soc.*, 1985, **107**, 4504.

Further Reading

E. L. Allred, J. Sonnenberg and S. Winstein, *J. Org. Chem.*, 1960, **25**, 26.
H. C. Brown and G. Zweifel, *J. Am. Chem. Soc.*, 1959, **81**, 247.
A. Pelter, K. Smith and H. C. Brown, *Borane Reagents*, Academic Press, New York, 1988, p. 165.

5

Chirality without Stereogenic Carbon

Aims

By the end of this chapter you should be familiar with:

- Appropriately substituted allenes and biphenyls as examples of compounds without a stereogenic centre
- The absolute configurations of chiral allenes and biphenyls
- Formation of chiral optically active compounds in which the stereogenic centre can be, for example, Si, Ge, Sn, N, P, As, S or Se

5.1   Allenes and Related Molecules

Nearly all the examples of chiral molecules so far encountered possess one or more stereogenic carbons. In this chapter the scope is widened, and organic molecules are considered that are chiral but do not have stereogenic centres, as also are some molecules in which the stereogenic centre is not carbon. In all cases the condition for chirality, namely that a molecule and its mirror image are not superimposable, is met.

The first class to be considered is the allenes. Allene (propa-1,2-diene) itself is represented in **1**, and as a Newman projection in **2**; allene is isomeric with propyne. The two hydrogens at each terminal carbon have a bond angle, HCH, of 120°, and of course these three atoms lie in a plane. However, in allene the two HCH planes are at right angles to each other, and are said to lie in **orthogonal** planes. This is apparent from the Newman projection **2**. It follows that the two π bonds are also orthogonal, and accordingly there is no conjugation between the π bonds. A representation of the bonding in allene is given in **3**; note that the central carbon is sp hybridized. Allenes are not alone in having π bonding of this type; other examples include ketenes, $R_2C=C=O$, isocyanates,

In allenes (a) the adjacent π bonds are orthogonal and (b) each terminal carbon and its two substituents lie in a plane, and these planes are orthogonal.

RN=C=O, diimides, RN=C=NR, the nitronium ion, $[O=N=O]^+$ and, most commonly, carbon dioxide, O=C=O.

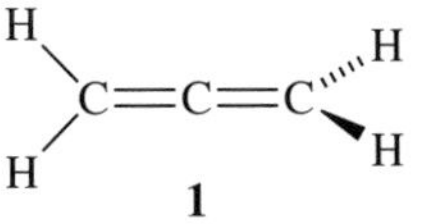

The prediction that certain allenes could exist as enantiomers was made by van't Hoff in 1875; Maitland and Mills finally verified his prediction (in 1935/36) after intense effort in many laboratories.

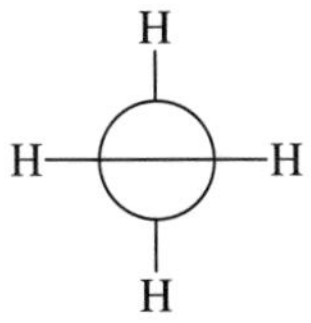

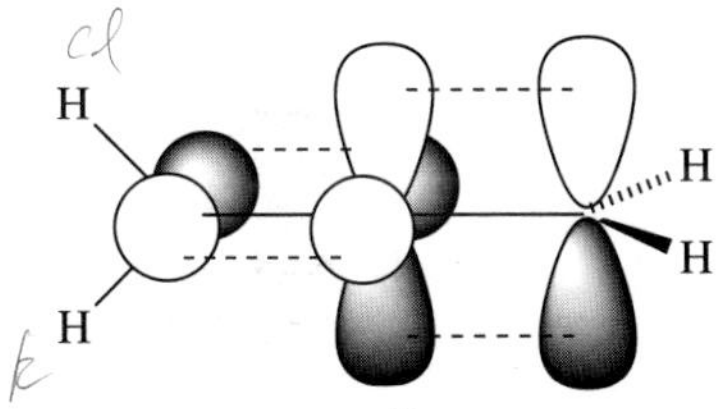

We consider the example of 1,3-dichloropropa-1,2-diene, which is shown with its mirror image in **4a** and **4b**. These two structures are related as enantiomers.

> If possible make a model to verify this, though not all model kits will have the facility to represent the central carbon. Take structure **4b** and rotate it by 180° around a vertical axis through the central carbon; this gives **4c**, and neither by this, nor by any other manoeuvre, can **4b** be superimposed on **4a**.

Enantiomers of chiral allenes are well documented and these molecules are not prone to racemization.

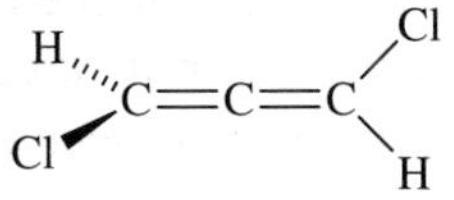

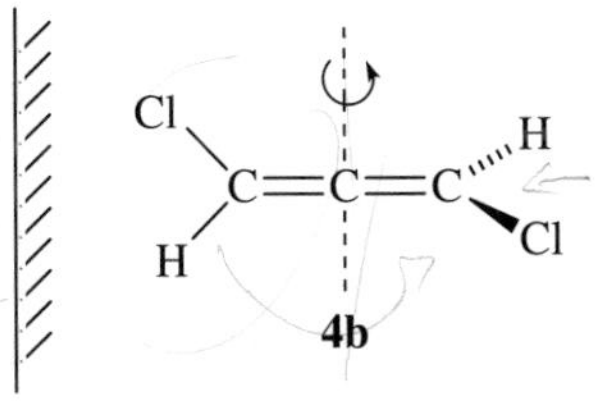

Although we have shown that **4** is chiral, and that it has no stereogenic centre, it is not without symmetry. This is best seen with the aid of a Newman projection of **4**, which is shown in **5**. A two-fold (C_2) axis of symmetry exists as shown; this passes through the central carbon, C(2), and bisects the right angle between the two chlorine atoms (and likewise the two hydrogen atoms). Rotation about this axis by 180° gives an identical molecule. Because of the axis of symmetry, **4** cannot be said to be asymmetric, and similar situations are found in certain other chiral molecules. This has had implications for stereochemical nomenclature. In particular, to avoid confusion the term 'asymmetric carbon' is now little used even for an sp^3 hybridized carbon that carries four

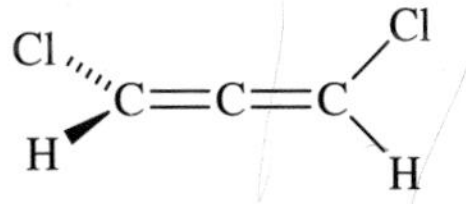

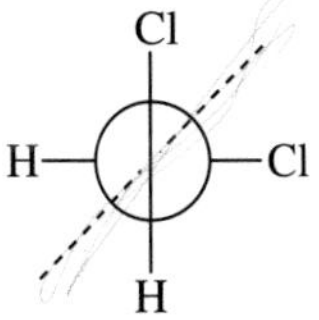

Allenes with the same two substituents at each terminal carbon are chiral, can be resolved, and are optically stable. Allenes with three or four different substituents are also chiral. These molecules *have no stereogenic centre.*

Alkylidenecyclohexanes and cyclohexanone oximes, substituted as in **6** and **8**, respectively, have no stereogenic centre and are also chiral.

different substituents, and the term has been superseded by 'stereogenic carbon'. The phrase 'asymmetric synthesis' survives, however. The demonstration that chirality can exist in a *molecule with an axis of symmetry*, e.g. **4**, is in direct contrast to the case of a *molecule with a plane of symmetry* (Chapter 3), in which chirality is not possible.

In order for an allene to be chiral a *minimum* of two different groups, both of which must be present at *each* terminal carbon, as in **4**, is required. This situation contrasts with the four different atoms or groups that must be bonded to an sp³ hybridized carbon to create a chiral molecule. However, chiral allenes may, of course, carry three, or four, different groups, and enantiomers of such examples are known.

Compounds such as the alkylidenecyclohexane **6** are distinct from, but related to, chiral allenes. Here, one of the double bonds of the allene is replaced by a six-membered ring in which the substituents at the 4-position are the same as those at the terminal vinyl (ethenyl) carbon.

> With the aid of a molecular model, verify that the resultant molecule **6** is chiral and note that the pattern of substituents in **6** is reminiscent of that in **4**, and that again there is no stereogenic carbon.

Compound **6**, as with disubstituted allenes such as **4**, illustrates the minimum requirements for chirality, but alkylidenecyclohexanes are known in which four different substituents occupy the key positions. Interestingly, the carboxylic acid **7** was the first example of this type of compound to be resolved into enantiomers, and this feat was achieved over 25 years before the resolution of an allene. One can, of course, consider other than six-membered rings in the context of chirality without stereogenic centres, and examples of alkylidenecyclobutanes are well documented (see Rossi and Diversi[1] and Runge[2]).

We have previously seen how geometric isomerism can exist around both the carbon–carbon double bond in an alkene and the carbon–nitrogen double bond in, for example, an oxime (Chapter 4), and we have now seen how an alkylidenecyclohexane such as **6** can be chiral without the need for a stereogenic carbon. Can one therefore obtain a chiral molecule in which the alkene part of **6** is replaced by, say, an oxime? Oxime **8** has four different groups in the key positions. The nature of the oxime group, in particular the lone pair on N, makes it impossible to construct a stable chiral oxime with only two different groups.

Make a model of the oxime **8** and also of its mirror image. Show that they are not superimposable, and note that indeed oximes such as **8**, that have no stereogenic centre, have been shown to exist as stable enantiomers.

5.2   Biphenyls

Another class of molecules that does not possess stereogenic centres, and which when appropriately substituted can be chiral and resolvable, is the biphenyls; biphenyl itself is shown in **9**. We consider biphenyls with only *ortho* substituents and these can be chiral if two conditions are met. It is tacitly assumed that the substituents do not themselves contribute to any chirality, *i.e.* they do not contain stereogenic centres. These two conditions are:

1. Resolvable biphenyls must each contain bulky *ortho* substituents of sufficient size that rotation around the single bond that connects the aromatic rings is prevented by the 'collision' of substituents. If rotation around the single bond does occur, then one conformation, that in which the two aromatic rings are co-planar as in **10**, has a plane of symmetry and this rules out chirality.
2. Resolvable biphenyls must contain two *different ortho* substituents on *each* ring (hydrogen can, of course, count as a substituent, though it is diminutive). If one, or both, rings contain two identical substituents, the molecule is not chiral. In order to make this point clear, consider the molecule **11** and make a model of it. In **11** the molecule is represented in a key conformation with the two aromatic rings mutually perpendicular.

The problems outlined in condition (2) can be overcome by interchanging one NO_2 group in **11** for a CO_2H group; this gives the isomeric molecule **12**.

Ortho-substituted biphenyls also do not possess a stereogenic centre, but exist as stable enantiomers if (a) the substituents are large enough *and* (b) the *ortho* sites on *each* ring contain different substituents.

With the aid of the model of **11**, verify:
(a) that there is a plane of symmetry that passes through the six carbons of the left-hand ring, and also through C(1) and C(4) of the right-hand ring;
(b) that there is a corresponding plane, at right angles to the plane described in (a), that passes through six carbons of the right-hand ring and through C(1) and C(4) of the left-hand ring.

Make a model of compound **12** and to your satisfaction demonstrate that the three-planes-of-symmetry obstacle to chirality, outlined in (1) and (2) and the exercises (a) and (b) above, has been overcome. It may be difficult to model a nitro group, but any sizeable alternative will do instead. Compound **12** was the first chiral substituted biphenyl to be resolved into enantiomers.

Appropriately substituted biphenyls are enantiomeric because of hindered rotation around a single bond, and are known as atropisomers.

The demonstration of chirality in **12**, and in other similarly substituted compounds, represents a new situation in that the chirality is caused by restricted rotation around a single bond [see condition (1) above], subject to the substituent patterns in the biphenyl [see condition (2) above]. The names given generally to isomers so derived are **atropisomers**, and to the phenomenon **atropisomerism**. 'Atropisomer' is derived from the Greek, 'without turning'. Hindrance to rotation at normal working temperatures may be very severe, in which case the atropisomers will be configurationally stable indefinitely. Alternatively, racemization may occur over time if the sizes of the relevant *ortho* substituents are such that they can slip past each other and permit rotation around the carbon–carbon single bond that connects the aromatic rings. If this happens, condition (1) above is no longer fulfilled, because there is a plane of symmetry in the conformation in which both aromatic rings are co-planar. Rates of racemization have been measured for certain combinations of substituents, in the case of **13**, with variable R; half-life times for the first-order racemizations are given in Table 5.1. As might be expected, there is a clear indication from the data in Table 5.1 that racemization is easier the less bulky the substituent.

An enantiomerically pure biphenyl will racemize if the *ortho* substituents are insufficiently large.

Table 5.1 Half-live(s) for racemization, $t_{1/2}$(rac), of substituted biphenyls **13**

R	T (°C)	$t_{1/2}$(rac) (min)
Me	118	179
NO$_2$	118	125
CO$_2$H	118	91
OMe	25	9.4

5.3   Absolute Configuration of Allenes and Biphenyls

5.3.1   Allenes

In order to describe the absolute configuration of chiral allenes, the Cahn–Ingold–Prelog (CIP) rules are first used to ascertain the priorities at the each terminal carbon, as is done for alkenes; at this point the similarity with alkenes ends. The method will be demonstrated with the trisubstituted allene **14a**:

1. Look along the three-carbon allene unit from *one end* (*either* end can be chosen), and put uppermost the higher priority group at the 'near' end. Initially we look from the left in this example, which already has the 'front' substituents in a vertical plane.
2. Represent **14a** as a projection **15**, with the 'near' substituents vertical, and the 'back' substituents horizontal.
3. The 'front' substituents, prioritized, take precedence over the 'back' substituents, themselves prioritized. This gives Br > Me; CO_2H > H.
4. The overall sequence is as shown in **16**. Track this sequence alphabetically, noting especially the angular direction taken in going from 'b' to 'c', *i.e.* from the lower priority front substituent to the higher priority back substituent. If the sense of turn is clockwise, the allene has R configuration, whereas if the turn is anti-clockwise the allene has S configuration. The enantiomer shown in **14a** therefore has S configuration.

Allene configurations are determined with the aid of the CIP convention. Prioritize the substituents at each end separately. Look down the allene from either end, such that the near higher priority group, a, is at the top; the near lower priority group, b, is at the bottom. Groups c and d are horizontal at the back. Track the groups alphabetically; a clockwise turn from b to c gives R configuration.

The last step is an adaptation of the original protocol (see Cahn[3]) and the method can also be conveniently applied to chiral biphenyls.

To assign configuration with the carboxylic acid group as a 'near' substituent, proceed as follows. Place the eye at the right-hand side of **14a** as drawn, and look toward and along the three-atom allene unit. Rotate the molecule clockwise around the axis defined by the three allenic carbons, in order to place the higher ranked 'near' substituent uppermost. The molecule now looks as in **14b**, with corresponding projection in **17**. Substituent ranking is as in **16**, and one again finds that the allene has S configuration.

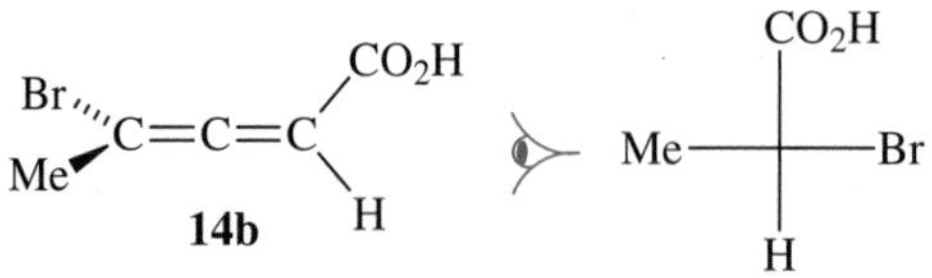

Just as the smallest carbocyclic ring that can incorporate a *trans* alkene is eight membered (Chapter 4), so the smallest sized ring that can contain an allene has nine carbons. One enantiomer of this compound is **18**. Rotate the molecule clockwise around the allene axis in order to place the higher ranked 'near' substituent uppermost. The projection from the right-hand side now has the form **19**, which is of general type **20**. Tracking the substituents alphabetically involves a clockwise turn between substituents 'b' and 'c'; hence the enantiomer of cyclonona-1,2-diene shown in **18** has *R* configuration.

> Configurations of biphenyls are determined the same way as for allenes; the *ortho* substituents must be included.

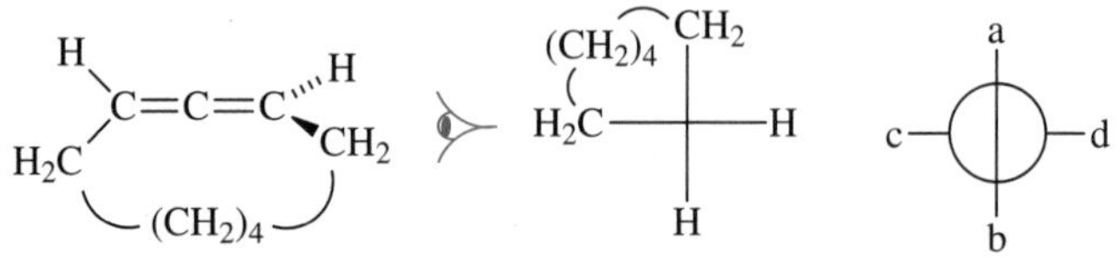

5.3.2 Biphenyls

Consider the case of the first chiral biphenyl to be resolved, one enantiomer of which is shown in **21**. To determine the configuration of this enantiomer, one draws a projection formula similar to that previously shown for allenes, but with inclusion of relevant *ortho* groups; this is shown in **22**. If one tracks the substituents alphabetically, then the sense of turn between 'b' and 'c' is anti-clockwise, as in **16**, and so **22** has *S* configuration. It is possible to construct molecules that additionally contain substituents at the unsubstituted aromatic carbons of **21**; however, any additional substituents are not relevant to the question of atropisomerism.

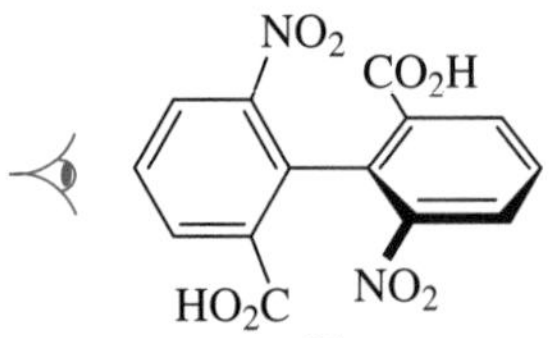

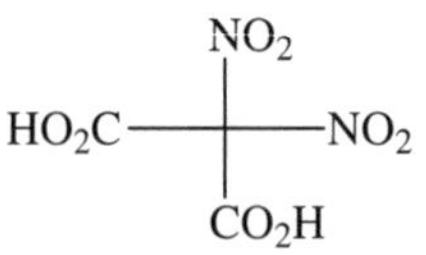

5.4   Hexahelicene

Hexahelicene (**23**) exists in two enantiomeric forms because of overcrowding, which affects the end two rings. The congestion is relieved by part of one of the end rings taking a position above the other so that the molecule starts to resemble the thread of a screw or helix. Hexahelicene can be resolved, either with the aid of a chiral chromatography column, or by formation of charge transfer complexes with a chiral complexing agent. The diastereomeric complexes are then separated by recrystallization, and very high values of the specific rotation are observed, $[\alpha]_D = 3700$ (note that this is not the *observed* rotation, but the specific rotation calculated as shown in Chapter 2).

23

5.5 Silicon, Germanium and Tin Compounds

A number of organosilicon compounds with chiral silicon atoms have been obtained as single enantiomers by routes involving:

1. Diastereoisomer formation and separation
2. Kinetic resolution
3. Asymmetric synthesis

Examples are shown in **24–26**. The stereospecificity of substitution reactions at silicon indicates a viable route to other enantiomerically pure silicon compounds. All examples of chiral molecules of the type $R^1R^2R^3R^4Si$ and $R^1R^2R^3SiH$ have high configurational stability, and in addition halides such as PhSi(Me)(Et)Br do not readily racemize (see Corriu *et al.*[4]).

(S)-**24** (S)-**25** (R)-**26** 1-Nap ≡ 1-Naphthyl

Single enantiomers of a small number of organogermanium compounds, with a chiral germanium atom, have been isolated and an example is the configurationally stable compound **27** (see Marshall and Jablonski[5]).

(R)-**27**   **28**

Tetraorganotin compounds with a stereogenic tin atom have been obtained as single enantiomers variously by resolution, asymmetric synthesis or, as in the example of **28**, separation of the enantiomers has been achieved by chiral chromatography on cellulose acetate. Compounds such as **28** exhibit no tendency to racemize, in contrast to those of general type $R^1R^2R^3SnOR$ in which an alkoxy group has replaced one organic ligand (see Davies[6]).

5.6   Amines, Ammonium Salts, Phosphorus and Arsenic Compounds

In amines, nitrogen inverts rapidly. Chiral amines can only be resolved if (a) the amine is part of a three-membered ring and adjacent to, *e.g.* oxygen, or (b) locked in a cage structure.

It is not generally possible to observe optical activity in amines of the type $NR^1R^2R^3$, in which groups R^1, R^2 and R^3 are independent and acyclic even though, with the lone pair on nitrogen classed as a formal substituent, all the requirements for existence of a stereogenic centre have been met. The problem is that the amino nitrogen is undergoing rapid pyramidal inversion (Scheme 5.1) and the energy barrier to inversion is usually small, about 25 kJ mol^{-1}. However, two factors diminish the rate of inversion of nitrogen in amines. Nitrogen has been shown to have configurational stability if it is part of a three-membered ring, and also to invert more slowly if an adjacent atom carries at least one lone pair of electrons, *e.g.* Cl, O or N. A number of chiral molecules that incorporate these features have been designed, and it has been possible to isolate both enantiomers of the chiral oxaziridine **29**, for example. These enantiomers do not racemize at room temperature (see Formi *et al.*[7]).

Scheme 5.1

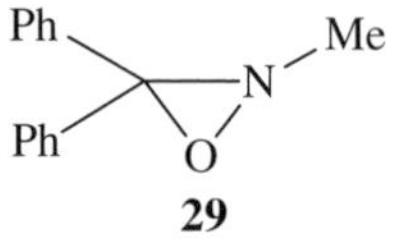

29 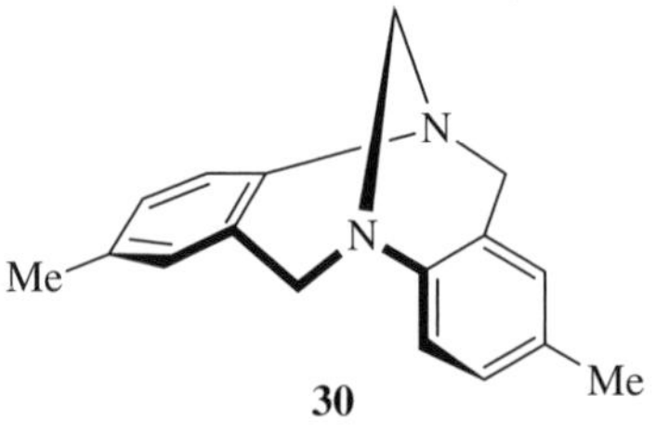   **30**

Chiral compounds have been synthesized in which nitrogen has been incorporated, as the only stereogenic centre, into a rigid cage structure, which prevents inversion. One such example is **30**, Tröger's base, which has two chiral nitrogens of interdependent configuration, since they are at the bridgehead positions of a bicyclic part structure. In one of the very first resolutions by chiral column chromatography, Tröger's base was separated into its enantiomers by means of the chiral disaccharide lactose.

Make a model of **30** and, treating the nitrogen in the same way as carbon, verify that the configuration at both nitrogen atoms is *S*.

Brucine (Chapter 3) and quinine (**31**), one of the first anti-malarial drugs and isolated from the bark of the cinchona tree in South America, are examples of naturally occurring alkaloids that contain a stereogenic nitrogen; additionally, these molecules contain a number of stereogenic carbons.

The enantiomers of quaternary ammonium salts such as $R^1R^2R^3R^4N^+X^-$ are isolable.

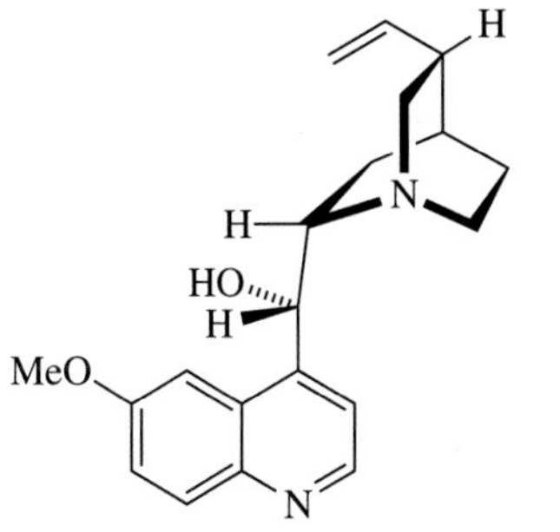

31

Identify the stereogenic carbons in brucine and quinine.

A quaternary ammonium salt has a tetrahedral nitrogen and if, as with carbon, this carries four different groups, it is chiral. A number of such salts have been resolved into their enantiomers, the earliest example being **32**. In general, these salts do not racemize, although reversible dissociation (Scheme 5.2) has been used to explain the few cases in which racemization does occur. The tertiary amine $NR^1R^2R^3$ so formed inverts configuration rapidly, and recombination with R^4X produces a racemic salt. The less nucleophilic the counterion X^-, the more the quaternary salt is configurationally stable.

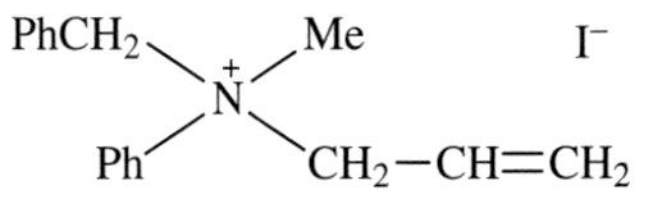

32

$$R^1R^2R^3R^4\overset{+}{N} \; X^- \;\rightleftharpoons\; R^1R^2R^3N \; + \; R^4X$$

Scheme 5.2

Single enantiomers of a number of chiral phosphines, **33** and **34**, have been obtained by reaction of the corresponding phosphine oxides with halosilanes. These compounds do not racemize readily, and in accord with this finding the barrier to inversion has been found to be much greater in the case of PH_3 (and also AsH_3) than in that of NH_3. Phosphonium salts, such as **35**, have been obtained by classical resolution routes. It is possible to abstract a benzylic proton from **35** to give an ylide that reacts with benzaldehyde to produce **36**, as a supposed intermediate. This sequence, the well-documented Wittig reaction, is completed by breakdown of **36** into the phosphine oxide **37**, with retention of configuration at phosphorus, and provides a convenient route to enantiomerically pure phosphine oxides.

Quaternary phosphonium salts such as $R^1R^2R^3R^4P^+X^-$, phosphines such as $R^1R^2R^3P$ and phosphine oxides of type $R^1R^2R^3P=O$ have all been obtained as single enantiomers that are configurationally stable.

A second route to phosphine oxides consists of a reaction between a diastereomerically enriched menthyl phosphinate **38**. This reaction proceeds with inversion of configuration at phosphorus, even though in

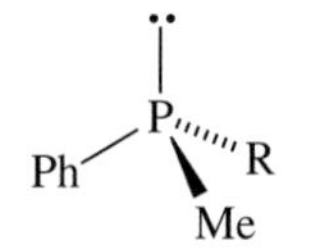

33 R = CH$_2$Ph
34 R = Bu^t

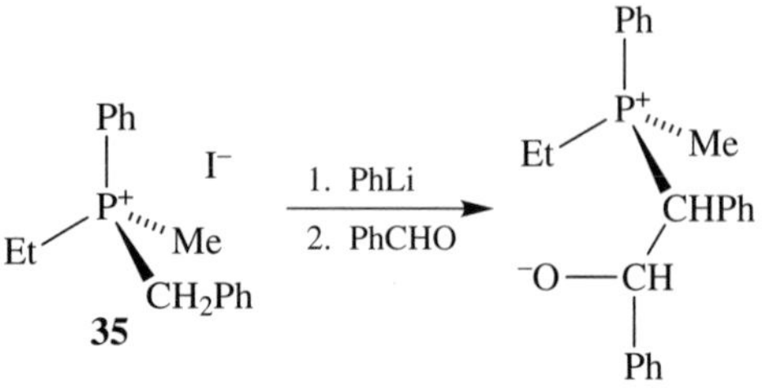

35

1. PhLi
2. PhCHO

36

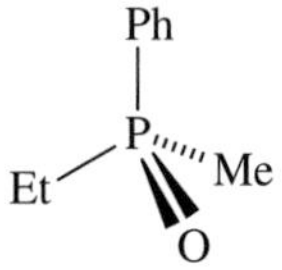

37

this example the CIP symbol for the phosphine oxide **39** is the same as that for the starting ester **38** (see Korpium et al.[8]).

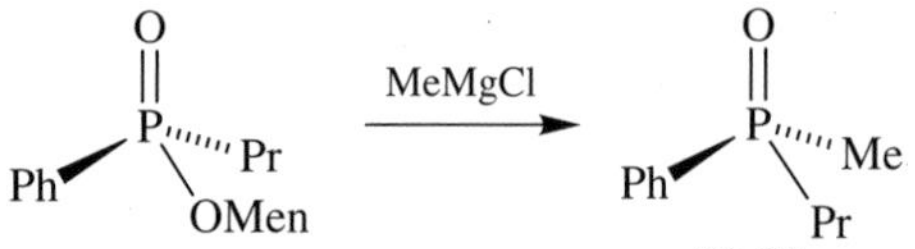

MeMgCl

(S)-**38**
Men = menthyl

(S)-**39**

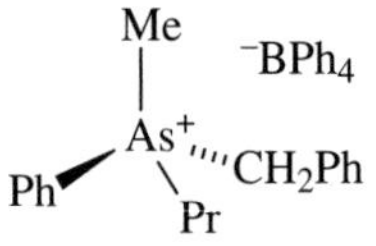

40

The first enantiomerically pure quaternary arsonium salts were isolated in the 1960s, and an example is shown in **40**. Importantly, the counterion here is the non-nucleophilic $^-$BPh$_4$. Earlier attempts to isolate salts that carried the nucleophilic counterion Br$^-$ led to racemization, presumably by a route analogous to that shown by Scheme 5.2 above.

Acyclic arsines have been obtained in enantiomerically pure form and, on account of their high inversion barrier, these compounds hold configuration indefinitely. We illustrate two methods for their formation from enantiomerically pure arsonium salts, and these are shown for **41** and **43**. Salt **41** is converted into **42** by cathodic reduction, and **43** is transformed into **44** by reaction with aqueous cyanide ion; also formed in the latter case is **45**, which accounts for the allyl (prop-2-enyl) group (see Wild[9]).

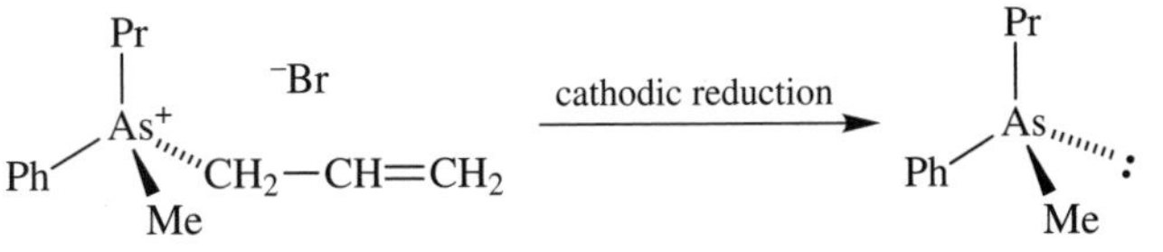

(R)-41 → (cathodic reduction) → (R)-42

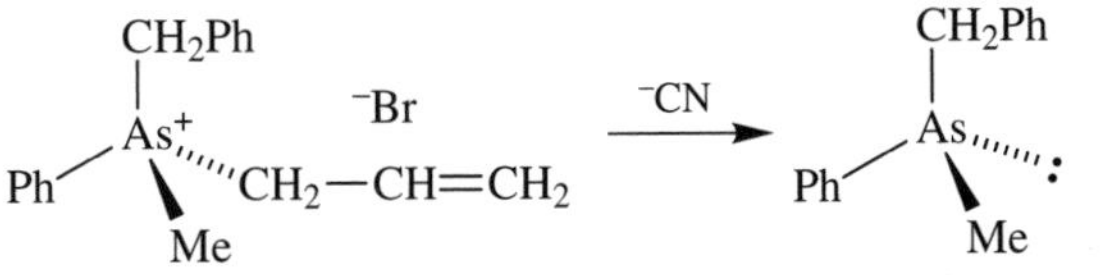

(S)-43 → (⁻CN) → 44 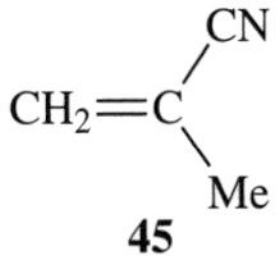 + 45

5.7 Sulfoxides, Sulfonium Salts and Selenoxides

Sulfoxides have a tetrahedral structure about the sulfur atom in which there are two alkyl and/or aryl substituents, an oxygen and a lone pair. The lone pair in sulfoxides does not undergo inversion, unlike its counterpart in amines, and so sulfoxides are configurationally stable; several examples of enantiomerically pure sulfoxides, *e.g.* **46**, have been obtained.

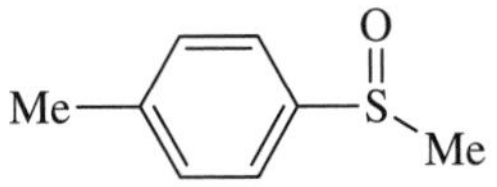

46

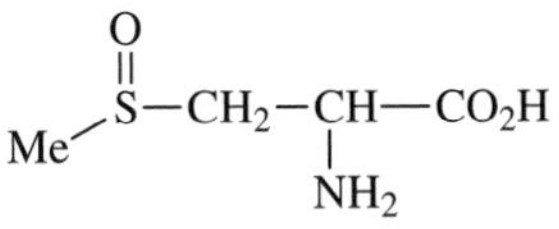

47

One example of a naturally occurring sulfoxide, available from turnips, is *S*-methylcysteine *S*-oxide (**47**); the CIP configuration at sulfur is (*S*). Draw a perspective diagram of this molecule.

Appropriately substituted sulfoxides, sulfinate esters and sulfonium salts exist as stable enantiomers. Some important examples exist in nature.

In sulfinic acids, RS(O)OH, sulfur is in the same oxidation state as in sulfoxides, and many esters of sulfinic acid (sulfinates) are known in which separate enantiomers have been isolated. If the ester is composed of a chiral alcohol, *e.g.* menthol, the resultant sulfinate is diastereomeric. Menthyl *p*-toluenesulfinate (**48**) has been used in synthesis.

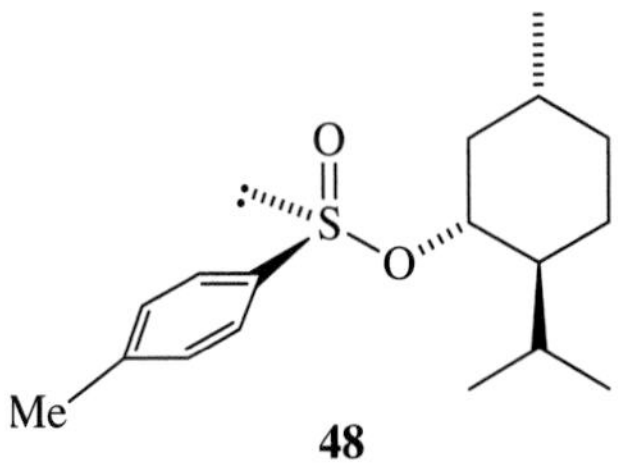

Sulfonium salts are of general form $R^1R^2R^3S^+X^-$. In these salts, positively charged sulfur has a lone pair of electrons, in addition to the substituent groups indicated. Sulfur is tetrahedral in sulfonium salts and salts of the above general type are accordingly chiral. Sulfonium sulfur does not undergo rapid inversion, and sulfonium salts have been

48

resolved; **49** is an example of an enantiomerically pure salt whose structure has been verified experimentally by X-ray analysis. In a few instances, ions such as **50** (episulfonium ions) have been isolated; these ions are intermediates in addition of sulfenyl halides (R–S–Hal) to alkenes, and as such have a generic similarity to bromonium ions (Chapter 4). One important sulfonium salt is *S*-methylmethionine (**51**), whose configuration is indicated. This compound functions as an important biological methylating agent, in which the methyl group can be transferred to, say, an oxygen atom.

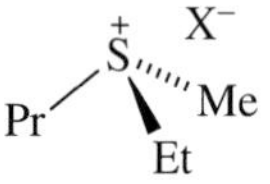

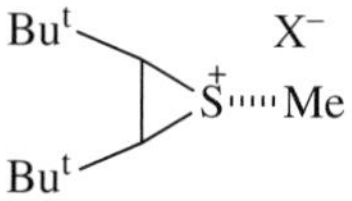

Enantiomerically pure sulfonium salts can undergo racemization, though the process is not usually rapid. Three processes have been implicated in racemization; these are: (1) reversible dissociation (Scheme 5.3a), similar to that suggested for the racemization of tetraalkylammonium salts (Section 5.6); (2) pyramidal inversion at sulfur, analogous to nitrogen inversion in amines; and (3) reversible dissociation into a carbocation and a sulfide (Scheme 5.3b) (see Andersen[10]).

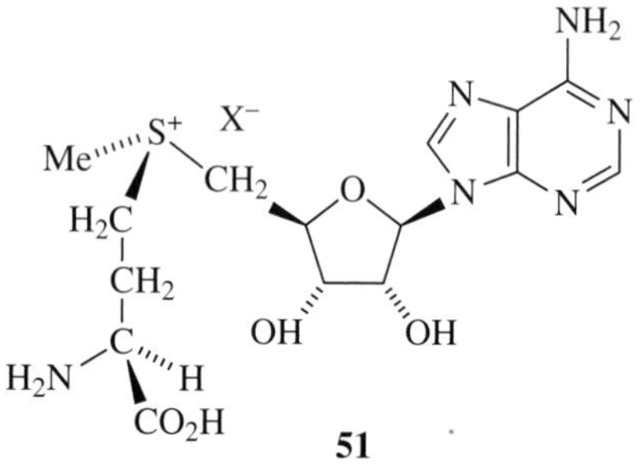

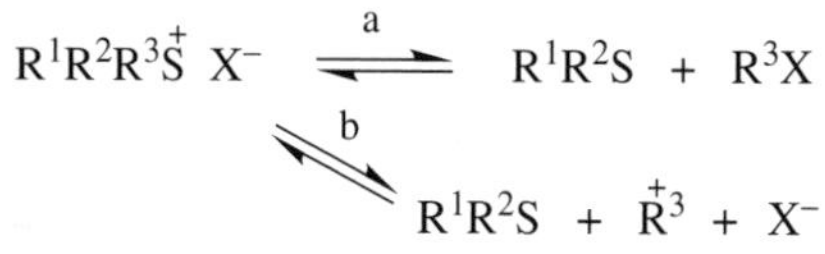

Scheme 5.3

In contrast to the corresponding sulfoxides, selenoxides of the type R^1R^2SeO are not usually chiral. The reason is that selenoxides react with water to produce structures of type **52** that are, of course, achiral. If the

Selenoxides are obtainable in enantiomerically pure form only in cases in which 'hydrate' formation is prevented.

$R^1R^2Se(OH)_2$
52

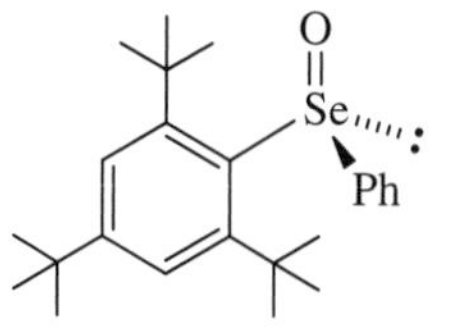

(*S*)-**53**

substituents are carefully chosen, a chiral selenoxide that does not racemize can be obtained. An example is **53**; the bulky substituents on one aromatic ring are sufficient to prevent attack by water at selenium that would lead to a 'hydrate' structure like **52** (see Shimizu and Kobayashi[11]).

Worked Problems

Q1. 2,2′-Binaphthol (**54**) is a chiral biaryl that is used in asymmetric synthesis. Determine the configuration of the enantiomer shown in **54**.

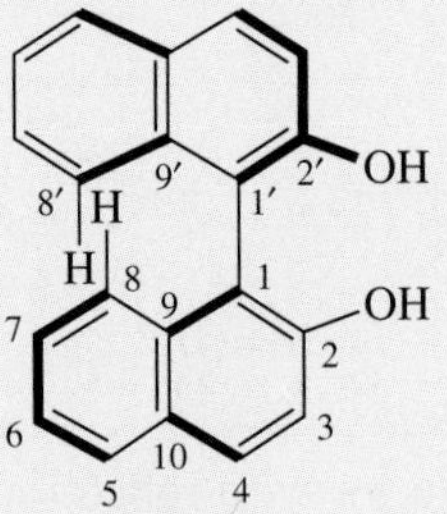

54

A. Compound **54** is chiral because of hindered rotation around the C(1)–C(1′) bond that connects the two naphthyl moieties. One of the causes of the hindered rotation is a repulsive interaction between H(8) and H(8′); although hydrogen is a small atom, molecule **54** is so constructed that these two hydrogens need to occupy exactly the same space for rotation to occur around the C(1)–C(1′) bond. The configuration of **54** is best determined using a molecular model, with particular reference to the carbons C(2) and C(9), and C(2′) and C(9′). One looks along the C(1)–C(1′) bond 'through' the lower right aromatic ring as drawn, and with the eye beyond C(4), and one places the higher priority near groups uppermost, and at the front, in a Newman projection. The 'back' groups are then placed horizontally. The resultant projection along the C(1)–C(1′) bond, with the eye beyond C(4) and looking along the C(1)–C(1′) bond, 'through' the lower right aromatic ring, is shown in **54a**. This projection is arranged in the same manner as the biphenyls and allenes with the higher priority 'near' substituent uppermost, and has the form **54b**. If one tracks the substituents from top front in the usual way, a clockwise turn is required as shown. The configuration of **54** is therefore *R*.

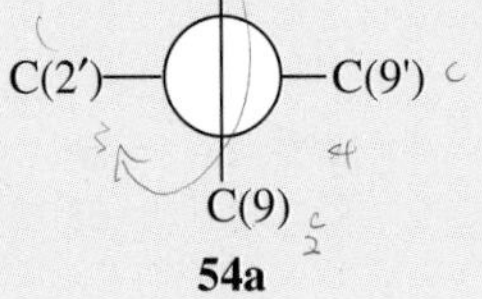

54a 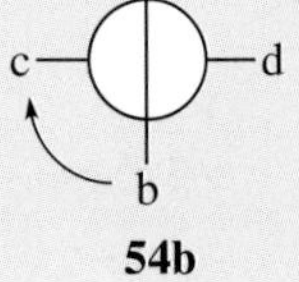   **54b**

Q2. An optically active biphenyl was synthesized in two isotopic modifications, the normal version **55** and the deuterio analogue **56**. These compounds undergo racemization in ethanol, and at $-19.8\ ^\circ$C the first-order rate constants for racemization were: for **55**, $k_{\mathrm{H}} = 6.48 \times 10^{-5}\ \mathrm{s}^{-1}$, and for **56**, $k_{\mathrm{D}} = 7.71 \times 10^{-5}\ \mathrm{s}^{-1}$. This gives a rate constant ratio $k_{\mathrm{H}}/k_{\mathrm{D}}$ of 0.84 (see Melander and Carter;[12] for related work see Mislow *et al.*[13]). What conclusions can be drawn concerning the relative sizes of deuterium and hydrogen?

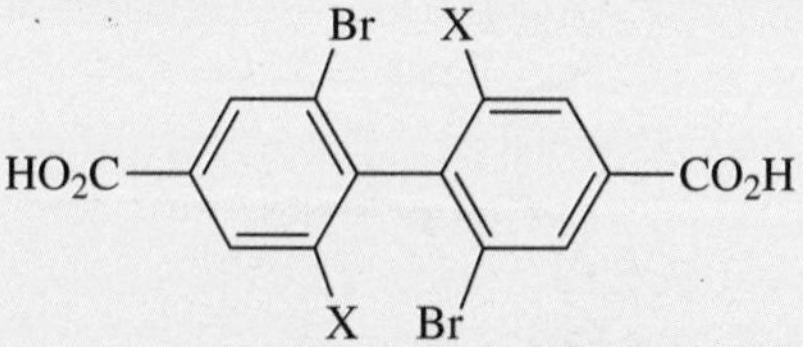

55 X = H
56 X = D

A. The relative rate constants are a reflection of the ease with which the bromine atoms in **55** and **56** pass the hydrogen and deuterium, respectively, and attain a planar conformation. On this basis, deuterium is smaller. Although initially surprising, this result is supported by other similar measurements, and is in accord with an earlier proposal that, since deuterium has a greater mass than hydrogen, the amplitude of vibration of a C–D bond is less than that of the C–H counterpart.

Q3. With reasons, and where possible with the aid of models, state whether you would expect compounds **57–63** to be chiral.

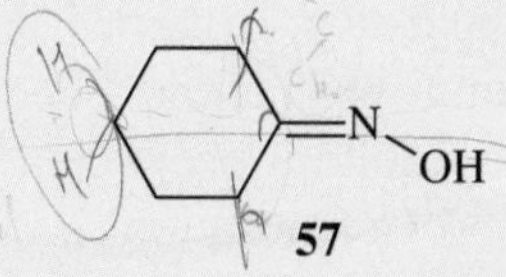

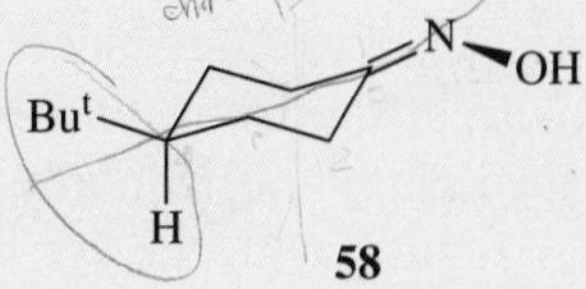

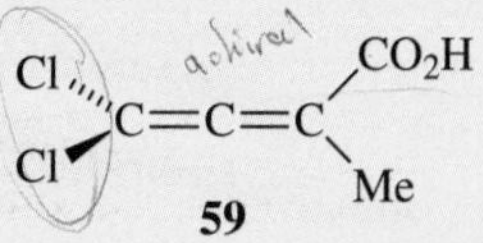

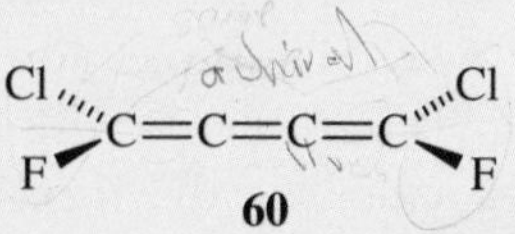

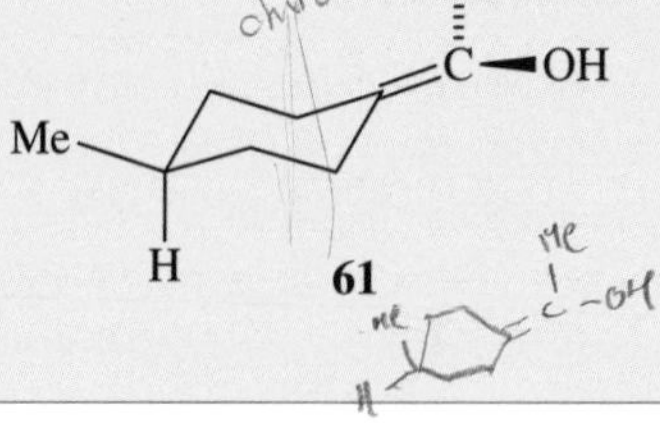

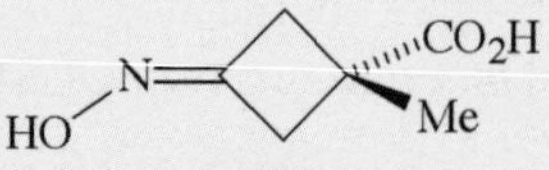

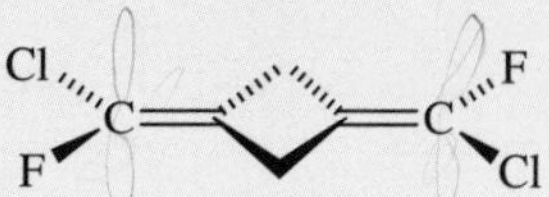

A. Compound **57**: no; the molecule does not possess substituents at C(4) of a six-membered ring.

Compound **58**: yes; the oxime and two different groups at C(4) make for chirality.

Compound **59**: no; the compound is of the form $C_{aa}=C=C_{bc}$ in which one terminal carbon is bonded to two identical substituents.

Compound **60**: no; this molecule contains three adjacent π bonds. The terminal π bonds are parallel to each other, and both are orthogonal to the central π bond. The substituents on the terminal carbons of the triene are now in the same plane. The situation is similar to that of an alkene, in that a change of configuration at one terminal carbon will produce a geometric isomer. The configuration of **60** is Z. (cis)

Compound **61**: yes; this compound is chiral, but remember that **61** is an enol and is readily converted to the more stable ketone **64**, which is not chiral. Any reversion to **61** from the ketone by base-catalysed enolization *via* the enolate, or acid-catalysed enolization, will produce both **61** and its enantiomer.

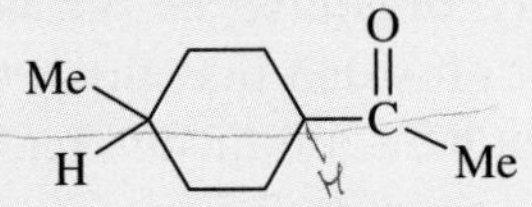

64

Compound **62**: yes; this is an oxime, and is likewise chiral. Unlike **61**, the oxime is the more stable tautomer, and under normal conditions shows no inclination to undergo a reaction analogous to ketonization.

Compound **63**: no; with respect to **60**, a four-membered ring has replaced the central double bond. A model will reveal that the chlorine and fluorine atoms are co-planar. Configuration is described with the E/Z nomenclature; **63** is E.

Summary of Key Points

- In this chapter we have considered three categories of chiral molecules that do not contain a stereogenic carbon atom, together with examples in which the stereogenic atom is not carbon.
- Allenes (with general substituents a, b, c, d) illustrate the important points that: (1) it is unnecessary for a molecule to have a stereogenic carbon for it to be chiral, and (2) an allene of the

type C_{ab}=C=C_{ab} possesses a two-fold (C_2) axis of symmetry and yet is chiral.

- An axis of symmetry in a molecule is thus not incompatible with chirality, in direct contrast to a plane of symmetry.
- Allenes with three different substituents arranged C_{ab}=C=C_{ac}, *e.g.* the substituent 'a' that occurs twice must be present at *both* terminal carbon atoms (see Worked Problems for an example of an allene C_{aa}=C=C_{bc}). Allenes of type C_{ab}=C=C_{cd} are also chiral, but now lack any symmetry axis.
- Compounds related conceptually to allenes, in which one double bond is replaced by a suitably substituted ring, *e.g.* 4-chlorocyclohexanone oxime, are also chiral, and again do not possess a stereogenic centre.
- *Ortho*-substituted biphenyls constitute a further distinct and synthetically useful class of compound that can be chiral provided that, in *each* ring, the *ortho* positions carry two different substituents (this pair of substituents *may* be present in *both* sets of *ortho* positions.
- The *ortho* sites may, of course, be substituted by: (1) three different substituents provided that the *ortho* substituent pattern is ab/ac and not aa/bc; (2) four different substituents.
- These restrictions on the choice of substituent ensure that planes of symmetry cannot exist when the two aromatic rings are orthogonal.
- If the *ortho* substituents are sufficiently bulky, rotation is prevented around the C–C single bond between the aromatic rings and this rules out a conformation in which the aromatic rings are co-planar, and which would possess a plane of symmetry.
- The enantiomers of *ortho*-substituted biphenyls are known as atropisomers.
- The R,S convention is applied, with modification, to determine the configuration of these three classes of compound.
- It is not generally possible to obtain enantiomers of amines that carry three different substituents because nitrogen undergoes a rapid and racemizing inversion.
- However, if nitrogen is adjacent to oxygen in a three-membered ring, inversion in the resultant oxaziridine is now retarded to such an extent that enantiomers can be isolated in suitably substituted cases (see Formi *et al.*[7]). Locking nitrogen at the bridgehead position in a caged ring, in order to prevent inversion, enables enantiomers to be isolated.
- Enantiomers of ammonium salts, $R^1R^2R^3R^4N^+X^-$, and sulfo-

nium salts, $R^1R^2R^3S^+X^-$ (sulfur also has a lone pair of electrons in these salts), are well documented. These can racemize under certain conditions, most commonly *via* a dealkylation–alkylation sequence.

- Sulfoxides of type R^1R^2SO and sulfinic esters such as $R^1S(O)OR^2$ have been isolated as stable enantiomers and the latter are useful in synthesis.
- Chiral selenoxides can be isolated where the substituents are sufficiently bulky to prevent the ready hydration, which gives an achiral compound.

Problems

5.1. With reasons, state whether each of the compounds **65–76** is chiral.

65

66

67

68

69

70

71

72

73

74

75

76

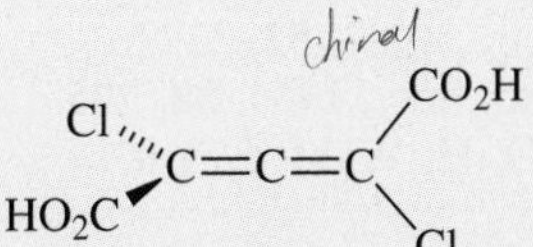

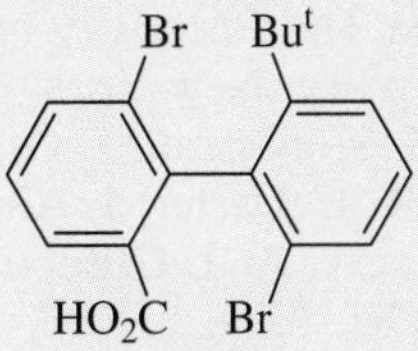

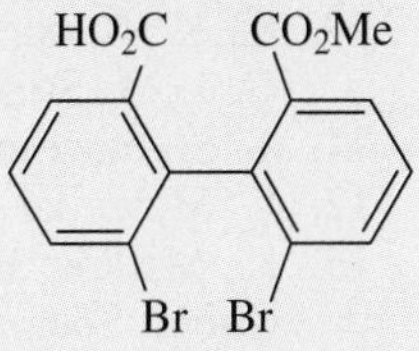

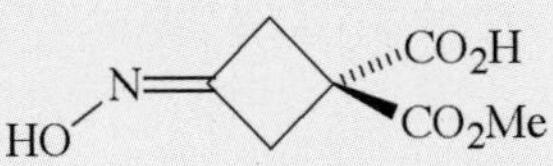

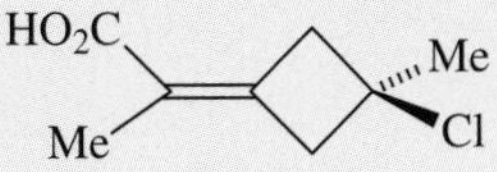

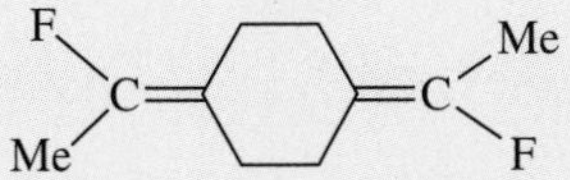

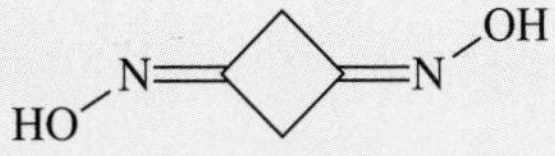

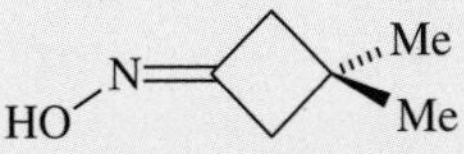

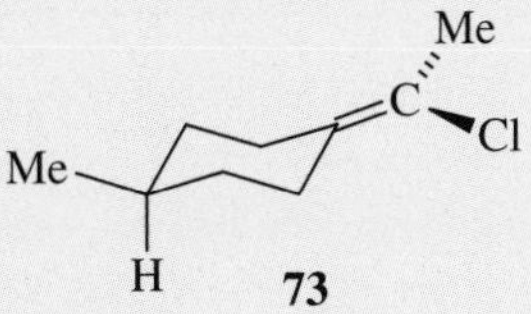

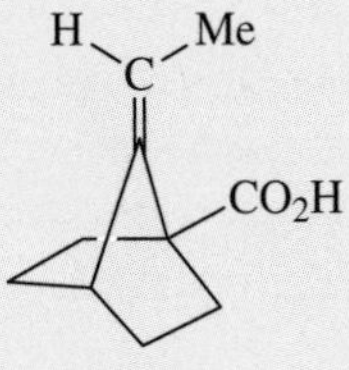

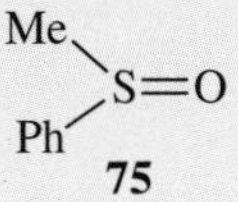

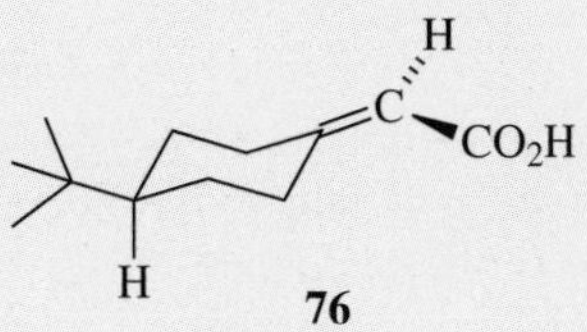

References

1. R. P. Rossi and P. Diversi, Synthesis, 1973, 25.
2. W. Runge, in Ketenes, Allenes and Related Compounds, ed. S. Patai, Wiley, New York, 1980, p. 99.
3. R. S. Cahn, J. Chem. Educ., 1964, 116.
4. R. J. P. Corriu, C. Guerin and J. J. E. Moreau, in Chemistry of Organic Silicon Compounds, ed. S. Patai and Z. Rappoport, Wiley, New York, 1989, p. 305.
5. J. A. Marshall and J. A. Jablonski, Chemistry of Organic Germanium, Tin and Lead Compounds, ed. S. Patai, Wiley, New York, 1995, p. 196.
6. A. G. Davies, Organotin Chemistry, VCH, Weinheim, Germany, 1997, p. 49.
7. A. Formi, I. Moretti and G. Torre, J. Chem. Soc., Chem. Commun., 1977, 731.
8. O. Korpium, R. A. Lewis, J. Chickos and K. Mislow, J. Am. Chem. Soc., 1968, **90**, 4842.
9. S. B. Wild, The Chemistry of Organic Arsenic, Antimony and Bismuth Compounds, ed. S. Patai, Wiley, New York, 1994, p. 89.
10. K. K. Andersen, in The Chemistry of the Sulphonium Group, ed. C. J. M. Stirling, Wiley, New York, 1981, p. 229.
11. T. Shimizu and M. Kobayashi, J. Org. Chem., 1987, **52**, 3399.
12. L. Melander and R. E. Carter, J. Am. Chem. Soc., 1964, **88**, 295.
13. K. Mislow, R. Graeve, A. J. Gordon and G. H. Wahl, J. Am. Chem. Soc., 1963, **85**, 1199.

6

Stereoisomerism in Cyclic Structures

Aims

By the end of this chapter you should be familiar with:

- *Cis* and *trans* 1,2-, 1,3- and 1,4-disubstituted cyclohexanes, and decalins, and with the use of dihedral angles of 0° and 180° in assignment of configuration
- Strain in cyclopropane and cyclobutane, and the much lower strain in cyclopentane and cyclohexane
- The *meso* nature of *cis*-1,2-di-X-substituted cyclopropanes, cyclobutanes, cyclopentanes and cyclohexanes
- The chiral nature of *trans*-1,2-disubstituted cyclopropanes, cyclobutanes, cyclopentanes and cyclohexanes
- The *meso* nature of *cis*-1,3-di-X-substituted cyclopentanes and cyclohexanes
- The achiral nature of the diastereomeric *cis*- and *trans*-1,4-di-substituted cyclohexanes
- The nature of steroids which contain mainly *trans*-fused cyclo-hexane rings
- The anomeric effect, which implies that there is a more exten-sive population of axial conformers in, for example, sugars than is suggested by the corresponding cyclohexanes

6.1 Cyclic Molecules, Configurational Assignment and Strain

Acyclic molecules can readily minimize strain simply by changing conformation. There is less freedom to make large changes of confor-mation in cyclic molecules and so the incidence of strain is much more widespread. Nearly all cyclic molecules possess some strain, though the amounts can often be small.

Strain has been encountered previously, *e.g.* in Chapter 1, and it is appropriate to collate the three types since cyclic molecules that are strained will experience one or more of the types outlined below.

1. Angle strain (formerly known as Baeyer strain) occurs when C–C–C single bond angles deviate significantly from the ideal value of 109°28′. These deviations occur either as a compression, or an expansion, of this angle.

2. Torsion strain (formerly known as Pitzer strain) is introduced when two substituents X and Y, bonded to adjacent carbons C(1) and C(2), are such that the dihedral angle X–C(1)–C(2)–Y is less than 60°. Torsion strain reaches a maximum when the above dihedral angle is 0°.

3. Steric strain. Two given atoms X and Y have specific van der Waals' radii, and if X and Y approach closer than the sum of their van der Waals' radii, a non-bonded repulsion is set up with the introduction of steric strain. If possible the molecule will respond by adopting a conformation in which the X and Y are further apart than the sum of their van der Waals' radii.

In disubstituted cyclopropanes, cyclobutanes and cyclopentanes, assignment of configuration, *cis* or *trans*, is made with respect to a plane through the molecule. Cyclopentanes and most cyclobutanes are not planar, but this does not pose a significant problem. Cyclohexanes are also non-planar. Using the above method it can take a while to decide whether the relative configurations of some disubstituted cyclohexanes are *cis* or *trans*. For these compounds we employ an alternative method based on dihedral angles.

6.2   Cyclopropane

Cyclopropane, C_3H_6, b.p. –33 °C, is the simplest cyclic hydrocarbon and was used as an anaesthetic. The C–C–C bond angles of cyclopropane (**1**) are 60°, much smaller than the ideal tetrahedral value of 109°28′. The three carbons of cyclopropane necessarily lie in a plane, and all the C–H bonds are eclipsed.

Cyclopropane possesses both significant angle and torsion strain. Strain in saturated cyclic hydrocarbons is determined from the heat of combustion, and comparisons are made per CH_2 group in a molecule. For cyclopropane the molar enthalpy change of combustion is 2090 kJ mol^{-1}, *i.e.* 696.5 kJ mol^{-1} per CH_2 group. The corresponding figure for cyclohexane is 3948 kJ mol^{-1}, *i.e.* 658 kJ mol^{-1} per CH_2. If cyclohexane is taken to be strain free (which is a reasonable approximation), one can conclude that the strain in cyclopropane is (696.5 – 658) = 38.5 kJ mol^{-1} per CH_2. It should be noted that these experimentally deter-

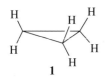

1

Cyclic molecules are frequently strained. Three types of strain, (1) angle strain, (2) torsion strain and (3) steric strain, are encountered.

mined heats of combustion represent enthalpies rather than free energies. A discussion of the hybridization and bonding of carbon in cyclopropane is given by Bernett.[1]

Cyclopropane can be drawn as in **1** to show perspective, or as in **2–5**
(for its derivatives), in which the cyclopropane ring is in the plane of the
paper. Monosubstituted cyclopropanes, such as **2**, are not chiral. The
disubstituted cyclopropane **3** is also not chiral; it has two stereogenic
centres with identical substituents and of opposite configuration, and is
therefore *meso*. Compound **3**, which has a plane of symmetry through
the mid-point of the C(1)–C(2) bond, is *cis*-1,2-dichlorocyclopropane.

Make a model of *cis*-2-chlorocyclopropane-1-carboxylic acid (a
compound related stereochemically to **3**) and its mirror image.
Decide whether or not this compound is chiral.

Cyclopropane has a very large
heat of combustion because it is
highly strained. Compounds such
as *trans*-1,2-dichlorocyclopropane
are chiral.

If, for example, there are two *trans* substituents, as in **4**, the molecule is
chiral. The configurations of the stereogenic centres of this *trans* compound are (1*R*,2*S*) and **4** is therefore (1*R*,2*S*)-2-chlorocyclopropanecar-
boxylic acid. A *trans* cyclopropane in which both substituents are the
same is also chiral, and an example is shown in **5**.

What are the configurations of C(1) and C(2) in **5**?

6.3   Cyclobutane

Cyclobutane (**6**) is not flat and one carbon is *ca.* 25° above the plane
defined by the other three. The internal CCC angles are 88°, a departure
of *ca.* 21° from the normal tetrahedral value. A planar structure would
have internal bond angles of 90° and eclipsed C–H bonds. Cyclobutane
has a strain energy of 27.5 kJ mol^{-1} per CH_2. The above values presumably represent an energy minimum and involve a slight increase in
angle strain and a slight decrease in torsion strain with respect to a planar structure. Certain substituted cyclobutanes, *e.g. trans*-cyclobutane-
1,3-dicarboxylic acid are, however, planar.

Nomenclature of cyclobutanes follows that of cyclopropanes.

Cyclobutane is not flat, though a
few of its derivatives are. *Cis*-
and *trans*-1,3-disubstituted
cyclobutanes are diastereomeric,
though neither is chiral.

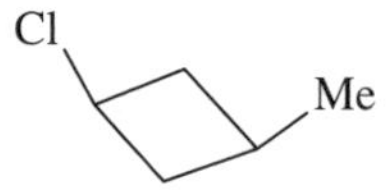

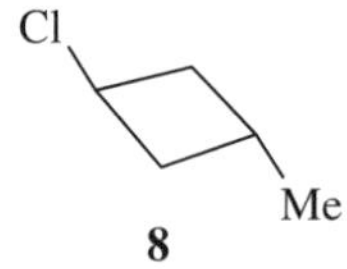

Compounds **7** and **8** (cyclobutanes are usually drawn in this style) are termed *cis* and *trans*, respectively. Further, **7** and **8** are diastereoisomers, though neither is chiral; this is significant in that diastereoisomers encountered previously, *e.g.* (*R*,*R*) and *meso* tartaric acids, have at least one chiral member. Indeed, this observation is general for disubstituted saturated cyclic hydrocarbons provided that the number of carbon atoms in the ring, *n*, is even and that the substituents are located at carbon atoms 1 and 1 + (*n*/2). For cyclobutanes this corresponds to C(1) and C(3), and for cyclohexanes C(1) and C(4). The stereochemistry of 1,2-disubstituted cyclobutanes is analogous to that of cyclopropanes.

6.4 Cyclopentane

Cyclopentane is appreciably less strained than cyclobutane and cyclo-propane, and the strain energy relative to cyclohexane is *ca.* 6.45 kJ mol^{-1} per CH_2 group. In order to lessen the torsion strain that would occur in a planar conformation, in which every C–H bond is involved in two eclipsing interactions, cyclopentane adopts a puckered conformation (see Dunitz, Further Reading). This has four carbons approximately planar, with the fifth carbon bent out of this plane in such a way that the mol-ecule resembles a small near-square envelope **9**. A Newman projection of **9** is shown in **10**.

Cyclopentane adopts an envelope, or puckered, conformation; the puckering rotates rapidly around the ring in a process called pseudorotation.

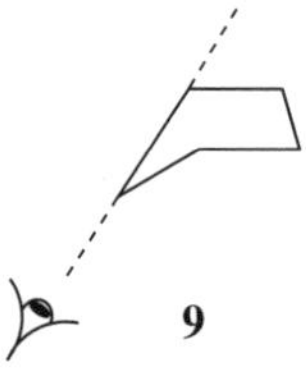

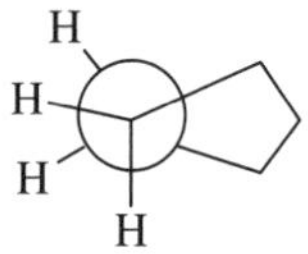

In this envelope-type conformation the puckering moves rapidly around the ring so that each carbon in rapid succession becomes the flap of the envelope. This process is known as pseudorotation (see Fuchs, Further Reading).

6.5 Cyclohexanes

In Chapter 1 the conformations of ethane, propane and butane were con-sidered and then extended to cyclohexane and its (more stable) chair and (less stable) boat conformations. This was done in order to show how the concepts applicable to acyclic molecules can be extended and applied to the most common cyclic saturated hydrocarbon. Cyclohexane has little strain and its derivatives, monocyclic or polycyclic, are abundant in nature. Indeed, this abundance is probably related to the approximately strain-free nature of cyclohexane. Whether cyclohexane is totally free of strain is a moot point, which has been addressed by Schleyer *et al.*[2] Cycloalkanes with rings consisting of seven to eleven carbon atoms are strained to the extent of 3.8–5.8 kJ mol^{-1} per CH_2, and in higher cycloalkanes the strain becomes very small.

Cyclohexane was originally postulated to be non-planar by the German chemist Sachse in 1890. Definitive proof was provided in *ca.* 1950 by Hassel in Oslo, from analysis of X-ray structures, and by Barton in London from correlations with steroid reactivities (see also Dunitz and Weser, Further Reading). The distinction between 'axial' and 'equatorial' positions was also made at that time (see Barton[3]).

In Chapter 1 we also considered how a bulky substituent prefers to be equatorial, and how (1) two hydrogens that are 1,2-diaxial are *trans* and (2) two hydrogens that are 1,3-diaxial are *cis*. In this chapter we develop these themes (see also Eliel and Wilen, Further Reading).

Cyclohexane is essentially strain free in monosubstituted cyclohexanes; the substituent has a greater tendency to be equatorial the bulkier it is.

6.5.1 Disubstituted Cyclohexanes

We now consider the chair conformations of cyclohexanes that carry two substituents and examine their stereochemistry and nomenclature. Where appropriate, chirality is considered. We deal firstly with cases in which *both* substituents can be axial, and then those in which *only one* substituent can be axial at any one time. Attention is focused principally on chair conformations.

6.5.2 Disubstituted Cyclohexanes in which Both Substituents can be Axial

1,2-Disubstituted Cyclohexanes

1,2-Disubstituted cyclohexanes are probably the most challenging of the disubstituted cyclohexanes. We consider 1,2-dimethylcyclohexane with both substituents axial in order to define configuration, although it should be recognized that it is the less stable chair conformation. Structure **11** represents the *trans* isomer because the dihedral angle C^α–C(1)–C(2)–C^β is 180°. The ring-inverted form **12**, arrived at from **11** solely by a series of interdependent bond rotations, and no bond cleavages, is therefore also *trans*. The chair conformation shown in **12** is the more stable because the methyl groups occupy equatorial positions where

trans-1,2-Di-X-cyclohexanes have two substituents either diaxial or diequatorial, and exist as a pair of enantiomers.

they experience only slight non-bonded interactions. One can also derive the configuration from the conformation shown in **12**, in which the tertiary hydrogens, H^1 and H^2, are bonded to C(1) and C(2), respectively. Now the dihedral angle H^1–C(1)–C(2)–H^2 is 180°, and so both chair conformations can be used to define *trans*-1,2-dimethylcyclohexane.

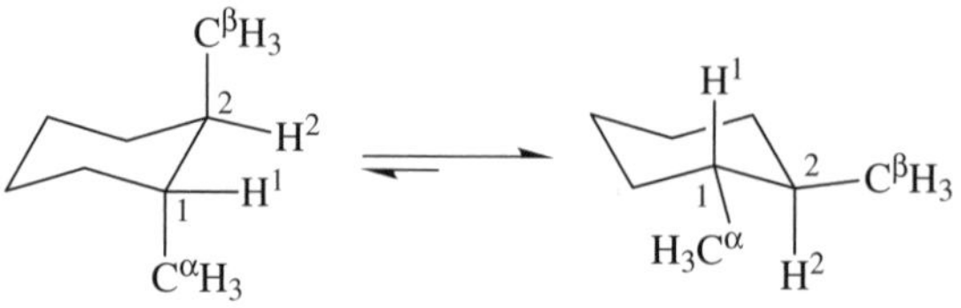

11   **12**

Make a model of **11** and of its mirror image **13** (which also has a diequatorial conformation **14**). Satisfy yourself that the pairs of structures **11** and **13**, and also **12** and **14**, are not superimposable by any conformational manipulation.

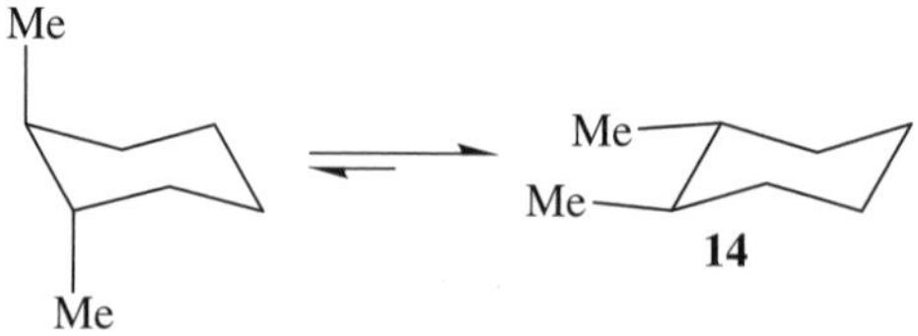

13   **14**

Accordingly, **11/12** and **13/14** are a pair of enantiomers; for a substituent X, one can make a general statement that *trans*-1,2-di-X-cyclohexanes exist as a pair of enantiomers.

1,3-Disubstituted Cyclohexanes

Here we consider 1,3-dimethylcyclohexane (**15**) and assign its stereochemistry. This compound has another chair conformation **16**, derived from **15** by ring inversion, and we now assign configuration from both conformers.

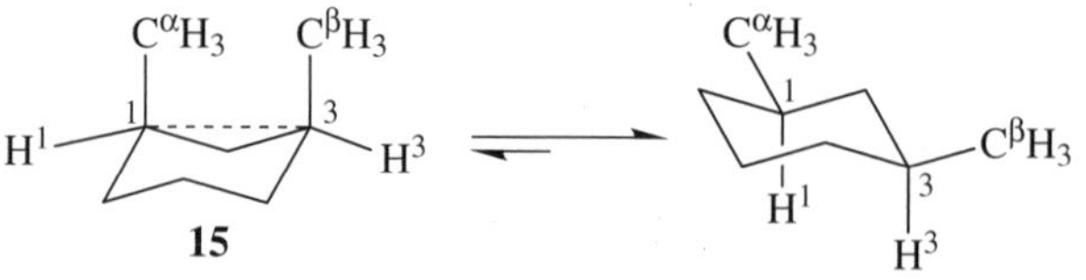

15   **16**

As in the previous sub-section, we search for a chair conformation in which there is a dihedral angle of 0° or 180°. Conformation **15**, with

diaxial methyl substituents, has a dihedral angle C^α–C(1)···C(3)–C^β of 0°
if one considers an imaginary bond between C(1) and C(3); according-
ly, **15** represents the *cis* configuration.

The ring-inverted chair conformation **16** is much more stable than **15**
because both methyl substituents are now equatorial. If one focuses on
the tertiary hydrogens H^1 and H^3 at C(1) and C(3), respectively, and
again considers an imaginary C(1)···C(3) bond, then in **16** the dihedral
angle H^1–C(1)···C(3)–H^3 is 0° and both conformations indicate that **15**
and **16** represent the *cis* configuration.

> Make a model of **15** and practise the ring inversion to **16**. Note
> that in either conformation one can readily identify a plane of sym-
> metry that passes through C(2) and C(5). Consequently, *cis*-1,3-
> dimethylcyclohexane is *meso*, and achiral, and this observation is
> general as long as the two substituents are identical.

cis-1,3-Di-X-cyclohexanes have
two substituents either diaxial or
diequatorial. These compounds
have *meso* configuration as long
as the substituents are identical.

> Make a model of both chair conformers of *cis*-1-chloro-3-methyl-
> cyclohexane. Is this compound chiral? Consider whether a plane of
> symmetry exists in any energetically accessible conformation.

1,4-Disubstituted Cyclohexanes

The chair form of a diaxially 1,4-disubstituted cyclohexane with two
identical substituents is shown in **17**. This pair of conformers is defined
as *trans* from either **17** or **18**, as follows. From **17** insert an imaginary
C(1)···C(4) bond and it is readily seen that the dihedral angle
C^α–C(1)···C(4)–C^β is 180°, in accord with a *trans* configuration.

Alternatively, take the diequatorial form **18** and, as before, insert an
imaginary C(1)···C(4) bond. One can now identify a dihedral angle
H^1–C(1)···C(4)–H^4 of 180°, which again indicates a *trans* configuration.

Conformations **17** and **18** together therefore represent *trans*-1,4-
dimethylcyclohexane. This compound has a plane of symmetry that
passes through C(1) and C(4), and also through C^α and C^β; according-

trans-1,4-Di-X-cyclohexanes have
two substituents either diaxial or
diequatorial. These compounds
are achiral because a plane of
symmetry passes through C(1)
and C(4).

ly, it is not chiral. Furthermore, it follows that no 1,4-*trans*-disubstituted cyclohexane is chiral as long as the substituents themselves do not induce chirality.

The configuration of disubstituted cyclohexanes considered so far can most readily be determined from analysis of those conformations in which the substituents (or alternatively the tertiary hydrogens at the carbons that carry the substituents) are diaxial. If the dihedral angles (normal, or extended in the 1,3- and 1,4-disubstituted cases) are 180°, the configuration is *trans*; if the dihedral angle is 0°, it is *cis*.

6.5.3 Disubstituted Cyclohexanes in which only One Substituent can be Axial

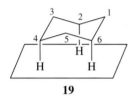

19

The cases of two identical substituents located, respectively, 1,2, 1,3 and 1,4 in chair cyclohexanes will now be addressed, in each case with relative configurations such that only one substituent can be axial at a time. At this point it is useful to consider cyclohexane **19**, resting on a horizontal surface by the axial hydrogens at C(2), C(4) and C(6); these three hydrogens lie in a reference plane. The axial C(1)–H bond is at right angles to the reference plane and directed away from it, whereas the equatorial C(1)–H is slightly, *ca.* 30°, inclined toward the plane. The axial C(2)–H bond is at right angles to, and directed toward, the plane and the equatorial C(2)–H bond is directed away from the reference plane by *ca.* 30°. With a molecular model of cyclohexane verify the above statements.

It follows, for example, that the axial C(1)–H and equatorial C(2)–H bonds, both being inclined away from the reference plane, are *cis*. The equatorial C(1)–H bond is inclined toward the reference plane whereas the equatorial C(2)–H bond is inclined away from the plane; the disposition of these bonds is therefore *trans*. In the present section we use two methods to assign the relative configuration of substituents. One is based on the angular inclination of bonds, described above, and the other derives from the relationship to examples, discussed previously, in which the substituents are diaxial.

1,2-Disubstituted Cyclohexanes

With one methyl group axial, and the other equatorial, we obtain **20** together with the inverted chair structure **21**. The structures here are equal in energy, and this is always the case when both substituents are identical. The compound **20** (or **21**) is described as *cis*, and this conclusion can be arrived at in two ways:

1.   A molecular model reveals that the bonds between the ring carbons and the attached methyl groups are inclined away from the plane

that passes through the hydrogens attached to C(1), C(3) and C(5) of **20**
[or C(2), C(4) and C(6) of **21**].

2. The structure is derived from the *trans* isomer by change of configuration at *one* carbon only, and the *trans* isomer has the 180° dihedral angle previously mentioned.

20   **21**

> Make a molecular model of both **20** and **21** and show that these are non-superimposable mirror images. Take **21** and invert the ring to re-obtain **20**. Conformational inversion has interconverted enantiomers.

cis-1,2-Di-X-cyclohexane has one X equatorial and the other X axial; this type of compound is *meso*.

Cis-1,2-dimethylcyclohexane represents an interesting case. The rapid rate of ring inversion means that it is impossible to separate the enantiomers. There is a further relevant point about *cis*-1,2-dimethylcyclohexane. The configurations of C(1) and C(2) in **20** and **21** are opposite, (1*S*,2*R*), and these configurations stay with the carbons whatever the conformation. Since the substituents at **20** and **21** are the same, *cis*-1,2-dimethylcyclohexane is *meso*. One can see this in another way: the boat conformation **22**, readily obtainable with models from either **20** or **21**, clearly has a plane of symmetry between the stereogenic carbons.

This assignment as a *meso* compound holds only for molecules in which both 1,2-ring substituents are the same. In this respect, *cis*-1,2-dimethylcyclohexane therefore follows the corresponding cyclopropane, cyclobutane and cyclopentane.

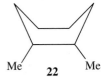

Me   **22**   Me

1,3-Disubstituted Cyclohexanes

With one substituent axial and one equatorial, **23** and its ring-inverted form **24** are obtained when the substituents are 1,3. For identical substituents the conformations are equal in energy. Moreover, **23** and **24**, being identical, are superimposable; however, this compound is chiral, and **23** and **24** together represent one of a pair of enantiomers.

trans-1,3-Di-X-cyclohexanes have one X equatorial and one X axial; compounds of this type exist as a pair of enantiomers.

23   **24**

Assignment of **23** (or **24**) as *trans* can be made as follows: (1) in the chair form of **23** (or **24**), one of the bonds between a ring carbon and a methyl group is inclined toward the reference plane defined at the start of this section, and the other is inclined away from it; (2) **23** (or **24**) is derived from **15** (which is *cis*) by a change of configuration at one ring carbon atom.

1,4-Disubstituted Cyclohexanes

These are represented by **25** and **26**, which are the two chair conformations of *cis*-1,4-dimethylcyclohexane, so called (1) because the bonds between the ring carbons and the attached methyl groups in **25** and **26** are both inclined toward the plane defined above, or (2) because **25** (or **26**) is derived from the *trans* diaxial isomer **17** by a change in configuration at one carbon.

25   26

cis-1,4-Di-X-cyclohexane has one X equatorial and one X axial. This compound is achiral because a plane of symmetry passes through C(1) and C(4). The *cis* compound and its *trans* counterpart are achiral diastereoisomers.

All *cis*-1,4-disubstituted cyclohexanes are achiral (providing that the substituents are not themselves chiral), because a plane of symmetry passes through C(1), C(4) and, of course, the substituents.

Cis-1,4-dimethylcyclohexane, **25** and **26** collectively, is therefore a diastereoisomer of its *trans*-1,4-dimethyl counterpart, **17** and **18**, although neither is chiral. This relationship therefore parallels that previously mentioned in respect of 1,3-disubstituted cyclobutanes (Section 6.3). The six configurational possibilities are summarized in Table 6.1.

Table 6.1 Configurational and conformational relationships in disubstituted cyclohexanes

Conformation of substituents	Relative configuration of substituents
1,2-Diaxial/1,2-diequatorial	*trans*
1,2-Axial-equatorial/1,2-equatorial-axial	*cis*
1,3-Axial-equatorial/1,3-equatorial-axial	*trans*
1,3-Diaxial/1,3-diequatorial	*cis*
1,4-Diaxial/1,4-diequatorial	*trans*
1,4-Axial-equatorial/1,4-equatorial-axial	*cis*

As revision, before proceeding, with the aid of models confirm that *cis*-1,2-dimethylcyclohexane and *cis*-1,3-dimethylcyclohexane are *meso*, and therefore achiral. Also make a model of both *trans*-1,2-dimethylcyclohexane and *trans*-1,3-dimethylcyclohexane and confirm that the molecules are chiral.

6.5.4 Other Disubstituted Cyclohexanes

If two substituents are not identical, the same principles of *cis/trans* nomenclature still apply. However, the degeneracy is removed in the case of *cis*-1,3- and *cis*-1,2-disubstituted compounds. For example, *cis*-3-fluorocyclohexanecarboxylic acid (**27**), shown in the more populated chair conformation, is now chiral, unlike the *meso* **15/16**.

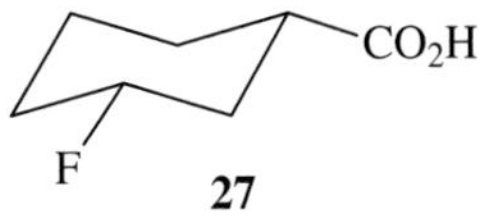

Make a model of **27** and its mirror image and confirm that these are enantiomers.

In monosubstituted cyclohexanes, the substituent prefers to be equatorial to an extent that is greater the larger the substituent. The substituent experiences destabilizing 1,3-diaxial interactions when axial, and so by ring inversion assumes the equatorial position (see Chapter 1). This inversion is an equilibrium process and the equilibrium ratios, of course, can be expressed as a free energy difference; in the present context they are known as '*A* values' and these are collected in Table 6.2.

The *A* values make it possible to obtain realistic estimates of relative axial:equatorial ratios in disubstituted cyclohexanes in which one substituent is axial and the other is equatorial. For example, from a knowledge of the preferences of the Me and OH groups to be equatorial one can assess approximately the conformational make-up, **28** *vs.* **29**, of *trans*-3-methylcyclohexanol.

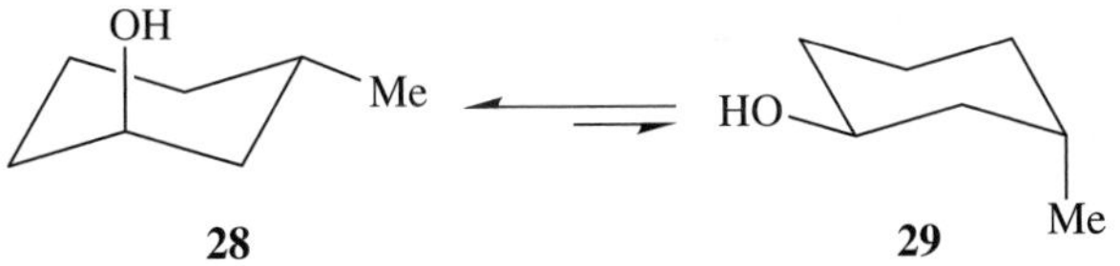

The *tert*-butyl (1,1-dimethylethyl) substituent is very bulky and exists solely in the equatorial position. The result is that the ring is unable to invert into the chair form that has this substituent axial, and so in effect the ring is locked. Winstein and Holness,[4] who first recognized this property of the *tert*-butyl group, noted that it offered 'a compelling but remote

Table 6.2 'A' Values of common substituents in cyclohexanes[a]

Substituent	A value (kJ mol⁻¹)	Substituent	A value (kJ mol⁻¹)
F	1.46	Bu^t	20.50
Cl	2.22	Ph	12.01
Br	2.22	$C{\equiv}CH$	1.72
I	2.05	$C{\equiv}N$	1.00
OH	3.77	$CH{=}CH_2$	6.23
OMe	2.64	COMe	4.27
OCOMe	2.97	CO_2Me	5.48
SMe	4.18	SO_2Me	10.46
NO_2	4.39	OSO_2Me	2.34
Me	6.70	NH_2	6.02
CF_3	10.46	MgBr	1.92
Pr^i	9.25		

[a]These A values are valid as a guide but, since they were measured in a variety of solvents, and at different temperatures, detailed comparison is not appropriate. For a comprehensive table of A values, including solvents used and experimental temperatures, see C. H. Bushweller in *Conformational Behavior of Six-Membered Rings*, ed. E. Juaristi, VCH, Weinheim, 1995, p. 25.

control of conformation'. Consider the *tert*-butyl group as part of a disubstituted cyclohexane, *e.g. trans*-1-*t*-butyl-3-methylcyclohexane. In principle, the methyl group can be axial and the *t*-butyl group equatorial, or *vice versa*.

Make a model of this compound and decide which is the minor, and which is the major, chair structure. Then carry out the same exercise with *cis*-1-*t*-butyl-2-chlorocyclohexane and *cis*-4-*t*-butylcyclohexanecarboxylic acid.

The assessments based on *A* values should be used with caution. Sometimes there is not a correspondence between experiment and calculation. For example, an excess population of the diaxial conformation of 1,3-diaxial cyclohexanes, over what is expected on the basis of individual *A* values, can be brought about by an attractive interaction, *e.g.* hydrogen bonding, between the substituents.

Occasionally, an apparent problem is encountered with the nomenclature of some 1,2-disubstituted cyclohexanes. This can be illustrated with **30** and **31**, *trans*-1,2-diethylcyclohexane. With a molecular model of **30** it is possible for the extremities of the two ethyl groups to touch, and so it may be intuitively difficult to accept that **30** is *trans*. However, the assignment of *trans* configuration is supported by: (i) the presence of tertiary diaxial hydrogens at C(1) and C(2) in **30**; (ii) observation that in the inverted conformer **31**, both ethyl groups are now 1,2-diaxial. We will return to this compound in Section 6.6.1.

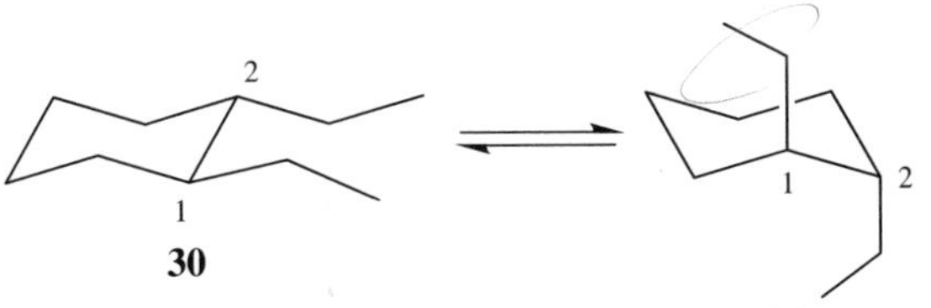

30

31

6.5.5 Trisubstituted Cyclohexanes

Cyclohexane may obviously carry more than two substituents, and the question arises of how to name these compounds. One may specify each centre (if it is stereogenic) by the *R/S* convention, or one may use the *cis/trans* system, described in previous sections. In the *cis/trans* system one substituent, the CO_2H group in the example **32**, is taken as a standard and given the prefix '*r*' (for reference), and all configurations in this molecule are given with respect to CO_2H. In this convention compound **32** is *trans*-3-fluoro-*cis*-2-methylcyclohexane-*r*-1-carboxylic acid. The *trans* relationship between the fluoro and methyl groups is not considered in this convention, but can be deduced from their relationship to the carboxylic acid. This convention, of course, gives no information about the absolute configuration.

In, for example, trisubstituted cyclohexanes, one substituent is taken as a reference, *r*, and the other two are denoted *cis* or *trans* with respect to the reference.

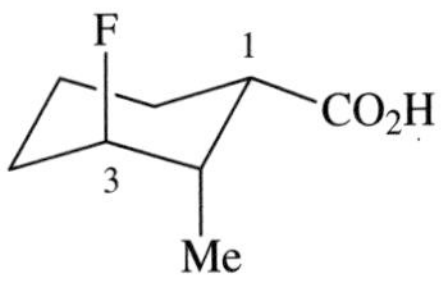

32

6.6   Decalins

The decahydronaphthalenes (decalins), $C_{10}H_{18}$, have two cyclohexane rings fused together with two carbons common to each ring. This fusion can be brought about in two ways and the resulting compounds are stereoisomers. They can be thought of as reduction products of naphthalene, though they are more commonly synthesized in other ways.

6.6.1 *trans*-Decalin

trans-Decalin is represented by **33**, or more commonly by **34**, and is seen to be *trans* in the following ways. In **34**, the vicinal ring-junction

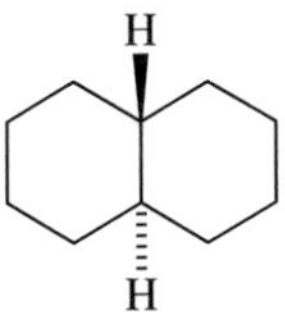

33

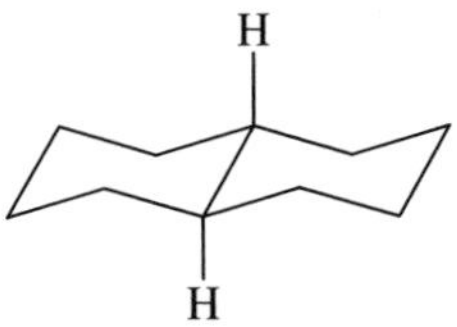

34

hydrogens are mutually *trans*. Consider also *trans*-1,2-diethylcyclohexane in the diequatorial conformation, as shown in **30**.

In the rigid *trans*-decalin the hydrogens at the ring junction are *trans* diaxial to each ring; also the second ring is formed by *trans* diequatorial CH$_2$ groups that link together.

> Make a molecular model of **30** and confirm that closure of the second ring gives **34**. After the ring closure, convince yourself that *trans*-decalin is *rigid around the ring junction*. It is possible only to flex the extremities of each six-membered ring in the model, to form mono- or di-boat conformations.

6.6.2 *cis*-Decalin

cis-Decalin is flexible and the rings can undergo inversion. With respect to each ring, the ring junction hydrogens are equatorial/axial and axial/equatorial. A similar analysis is possible using CH$_2$ groups adjacent to the ring junction.

The ring junction may form in a different way to give *cis*-decalin, which exists as two conformers **35** and **36**. These conformers are seen to be *cis* because the hydrogens H^a and H^b are *cis* to each other. Each is equatorial with respect to one ring, and axial with respect to the other.

H^a

H^b H^b

                    H^a

35

36

> Make a molecular model of *cis*-1,2-diethylcyclohexane (**37**) and form the second ring by ring closure. *cis*-Decalin in conformation **35** is formed. Satisfy yourself that, unlike its *trans* counterpart, *cis*-decalin is flexible around the central C–C bond. Then take the model of **35** and invert it into **36**; this may take a little practice as the rings do not maintain their chair structure in the models (nor indeed in reality) during inversion. Alternatively, take a model of *cis*-1,2-diethylcyclohexane, as shown in **38**, and close the second six-membered ring to give directly the conformer shown in **36**.

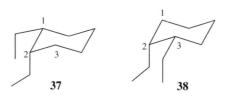

37   **38**

6.7   Steroids

Of course, it is possible to fuse together more than two cyclohexane (or related) rings. Nature does this very efficiently in the synthesis of steroids, which are widely distributed in the animal kingdom. Cholestanol (**39**) possesses four fused rings; three are six-membered and one is five-membered. In steroids such as **39** the rings are labelled A, B, C and D from the left, as drawn. In cholestanol, the A/B, B/C and C/D ring junctions are all *trans*. The two methyl groups at the ring junctions of **39** are termed 'angular methyl groups'. The closely related and important compound cholesterol (**40**), a necessary intermediate compound in the *in vivo* synthesis of steroid hormones, has a double bond in ring B; this induces a slight change in the geometry of cholesterol compared to **39**.

Steroids mostly have a *trans* A/B ring junction; in bile acids, however, this A/B ring junction is *cis*.

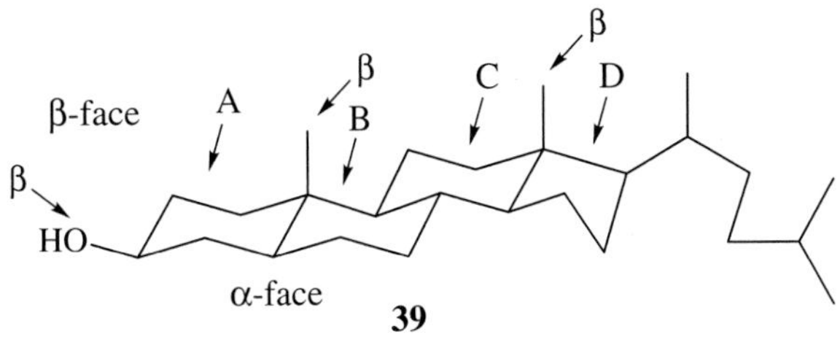

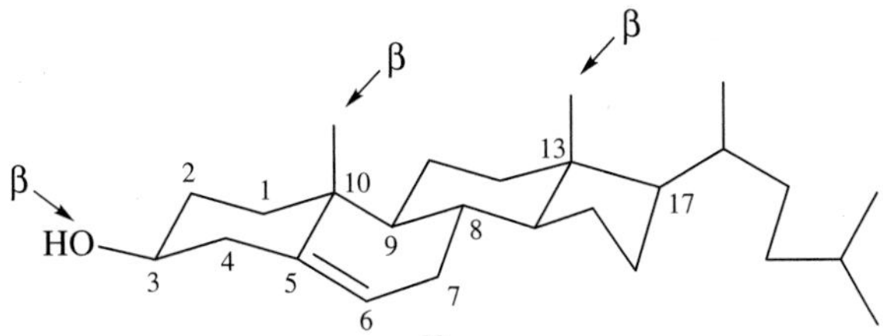

In steroids, groups that are *cis* to the angular methyl groups are above the plane of the molecule, as drawn, and are termed β, whereas those that are below the plane and are *trans* to the angular methyl groups are known as α. These designations are used in naming steroids; thus cholesterol (**40**) is also known as 5-cholesten-3β-ol.

Bile acids are found, as their name suggests, in bile, which is formed in the liver and then stored in the gall bladder. Of these acids, cholic acid (**41**) is the most abundant. A structural feature which cholic acid shares with the other bile acids is a *cis* A/B ring fusion, which is seen in the structure of **41**.

41

6.8   Anomeric Effect

We have discussed how substituents in chair cyclohexanes prefer to be equatorial rather than axial in order to avoid 1,3-diaxial interactions. The preferences are expressed as 'conformational free energy differences' or *A* values, and are mainly steric in origin. Certain sugars, such as the hexose D-(+)-glucose, exist in aqueous solution mainly as two cyclic structures, **42** (36%) and **44** (64%), which are interconverted *via* an open-chain structure **43** (0.02%) as shown in Scheme 6.1. Anomers **42** and **44** have been isolated separately, and have different melting points and specific rotations. In addition, both **42** and **44** exist as chair forms, similar to cyclohexane except that one ring member is now oxygen. The minor cyclic structure **42**, as drawn, has the OH group at C(1) axial and 'down'; in this structure the OH group at C(1) is *trans* to the CH_2OH group and **42** is given the symbol α. Similarly, in **44** the OH group at C(1) is 'up' and equatorial; it is *cis* to the CH_2OH group and this structure is given the symbol β. Structures **42** and **44** differ only in their configuration at C(1), which is the aldehyde carbon in the ring-open form **43**. Carbon C(1) is known as the **anomeric carbon** and **42** and **44** are known as α and β *anomers*, respectively. The anomeric carbon is easily recognized, as it is the only carbon bonded to two oxygen atoms. Compounds such as **42**, the simplest member of which is $MeOCH_2OH$, are called hemiacetals.

Both **42** and **44** have been isolated as separate compounds. The α-anomer **42** has $[\alpha]_D = +112.2$ and the β-anomer **44** has $[\alpha]_D = +18.7$. When *either* anomer is dissolved in water, it is slowly converted into the equilibrium mixture of anomers which has $[\alpha]_D = +52.6$. With this information it is possible to determine the ratios of α- and β-anomers at equilibrium. If x is the percentage of α-anomer, one can write: $x \times 112.2 + (100 - x) \times 18.7 = 100 \times 52.6$. Solving for x gives: $x = 36.2\%$. With the justified assumption that the concentration of the open-chain form is *ca.* 0%, it follows that the amounts of the cyclic forms present at equilibrium are: **42** (α-form) 36.2% and **44** (β-form) 63.8%.

As one of a number of such examples, the α-form of D-glucose (36%) is more abundant with respect to the β-form (64%) than expectations based on substituents in cyclohexane would indicate. This unexpected abundance of the α-form is known as the anomeric effect.

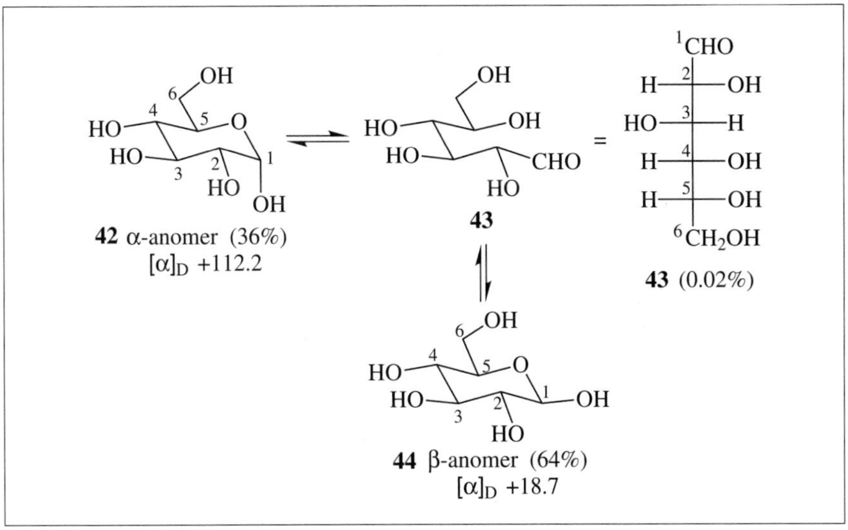

42 α-anomer (36%) [α]$_D$ +112.2

43

43 (0.02%)

44 β-anomer (64%) [α]$_D$ +18.7

Scheme 6.1

Similar behaviour, again mediated by an open-chain aldehyde form, has been observed with the related sugar D-mannose, in which the α-anomer **45** (69%) now predominates over the β-anomer **46** (31%).

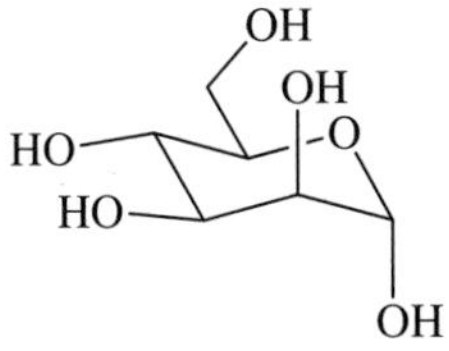

45 α-anomer (69%)

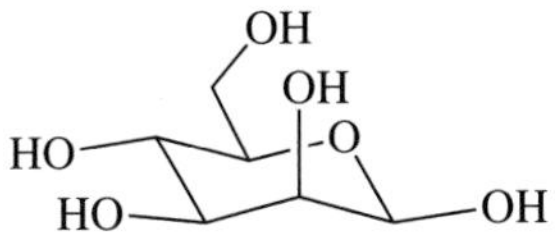

46 β-anomer (31%)

A further example is provided by the methyl mannosides **47** and **48**. In 1% methanolic HCl, equilibration of the anomers was achieved and the ratio α:β = 94:6 (see Senderowitz et al.[5]).

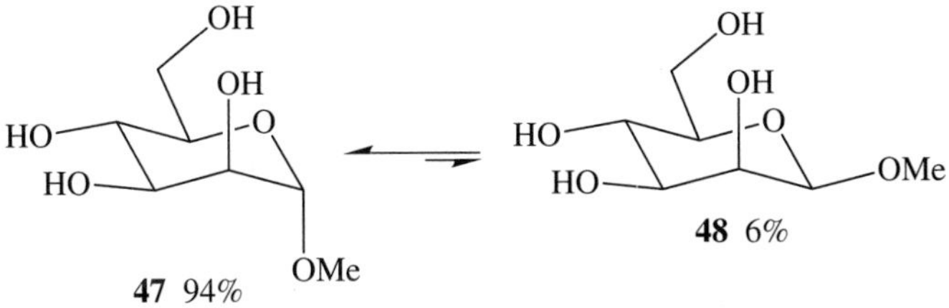

47 94%

48 6%

In these and other examples, the percentage occupancy of the axial position by OH, OMe, OCOMe and Cl significantly exceeds expectations based on the corresponding cyclohexyl derivatives (see Table 6.2 for a list of substituent A values). This excess population of the axial position, first observed with sugars, is known as the **anomeric effect**; it is exhibited only by *electronegative* substituents.

The anomeric effect is quoted as a free energy difference between axial and equatorial forms in tetrahydropyrans (THPs) with account taken of the greater steric preference of a substituent to be equatorial in a THP with respect to a cyclohexane.

If most of the ring substituents are removed from a sugar, a 2-X-substituted tetrahydropyran remains and the enhanced population of the conformation in which X is axial is observed here also, and is shown for the case of **49** and **50**. The anomeric effect is now expressed as a free-energy difference, which is derived from the equilibrium axial:equatorial ratios of X in, for example, tetrahydropyrans (THPs) **49** and **50**. Tetrahydropyrans are used in preference to cyclohexanes because the C–O bond (*ca.* 0.14 nm) in a THP is shorter than its C–C counterpart (0.154 nm) in a cyclohexane.

The shorter C–O bond length in the THP has the effect of increasing 1,3-diaxial repulsion. This means that, on steric grounds alone, the 2-X substituent in a THP would be equatorial to a greater extent than in a corresponding cyclohexane. Alkyl substituents, not being electronegative, do not exhibit an anomeric effect. These substituents show a greater preference for the equatorial position in THPs than in cyclohexanes. With the assumption that all substituents experience related pro-equatorial steric effects as alkyl substituents, it is therefore possible to calibrate the steric effects of the electronegative substituents. These steric effects are, of course, over-ridden by the opposing anomeric effect to give the observed axial preferences exhibited by THPs.

The relevant data are collected in Table 6.3. From the *A* value and the corresponding $\Delta G°$ value for the 2X-THP when X = Me, one can generalize this finding for Me and conclude that $\Delta G°(THP) = 1.5 \times A$ value for all substituents. It is then possible to deduce the 'real' value of the anomeric effect. From Table 6.3 one can see that the anomeric effect

Table 6.3 Anomeric effects (in kJ mol[-1]) of electronegative substituents in 2-X-tetrahydropyrans

Substituent (X)	A value	$\Delta G°(THP)$[a]	Anomeric effect
H	0	0	0
Me	7	12	0
OH	3.5	−0.5	4.5
OMe	2.5	−3.6	7.3
OCOMe	3	−2.0	6.5
Cl	2.5	−7.5	11.25

[a]Negative values for $\Delta G°(THP)$ indicate an axial preference

increases with greater substituent electronegativity. In THPs and sugars, therefore, a given substituent, X, exerts two opposing effects:

(1) A steric effect, which leads to a preferred equatorial population of X

(2) An 'electronegativity' effect, which increases the axial population of X

The anomeric effect is general in that broadly the same effects are shown, whether the saturated six-membered ring contains an oxygen, a nitrogen or a sulfur as the heteroatom. In the case of sulfur the C–S single bond (0.182 nm) is longer than its C–O counterpart, and so 1,3-diaxial interactions are less important in the thiacyclohexane case. Solutions of the trithiacyclohexane **51** do not contain any equatorial isomer.

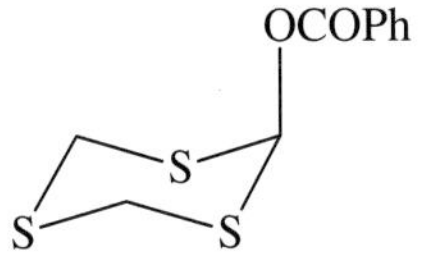

Two effects whose relative importance is uncertain cause the anomeric effect. These are (1) an n→σ* effect (Scheme 6.2) and (2) electrostatic repulsion.

Although the anomeric effect was first recognized with the hemiacetal group in certain sugars, one explanation is found in the behaviour of acetals. In general, as the R groups show, an acetal has a *gauche* conformation with respect to both C–O bonds as in **52**. Importantly, this places a lone pair on one oxygen, **antiperiplanar** to the bond between the 'central' carbon and the other oxygen, that is the dihedral angle lone pair–O–C–O in **53** is 180°. This conformation is known to be optimum for conjugative electron release (Scheme 6.2).

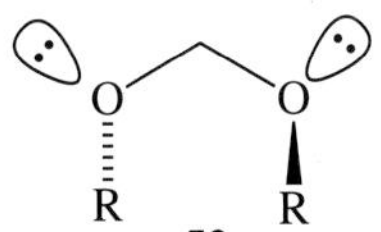

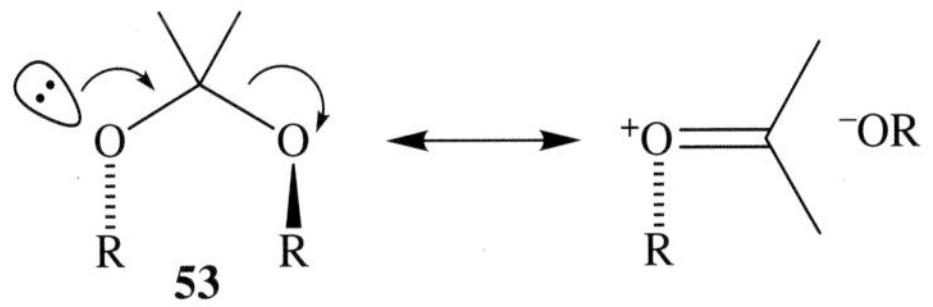

Scheme 6.2

One interpretation of the origin of the anomeric effect can therefore be considered to be a two-electron stabilizing shift (Scheme 6.2) from a lone-pair of electrons on one oxygen to the vacant σ* orbital of the adjacent C–O bond. Significantly, the antiperiplanar arrangement here (and elsewhere) permits maximum overlap of the relevant orbitals. The preferred *gauche* conformation (Scheme 6.2, left-hand structure) only exists in heterocycles if the group attached to the ring is axial (see Kirby[6]).

Verify this with the aid of a molecular model. Certain types of molecular models contain oxygen lone pairs. If such a model set is not available, use a carbon atom instead and paint (or mark) two hydrogen atoms to represent lone pairs. The bond lengths and angles at carbon are slightly, but not significantly, different from those at oxygen.

However, this interpretation is open to some doubt (see Box, Further Reading). One investigation that used compounds **54** and **55** gave results that are not in accord with the proposal above.

54   **55**

It is known that nitrogen is: (1) a very good conjugative electron donor (as shown for amides in Chapter 4) and certainly better than oxygen; (2) less electronegative than oxygen. One would expect from (1) that n→σ* interactions would be stronger in **54** than in **52**, and from (2) that dipole–dipole interactions would be weaker in **54**. No enhanced axial population was observed for **54** with respect to **55**. Values of the free energy difference between equatorial and axial conformations are 1.9 ± 0.2 kJ mol^{-1} for **54** and 2.6 ± 0.6 kJ mol^{-1} for **55**. The n→σ* effect, if it is present in **54**, is clearly not dominant.

When the substituent in **54** or **55** is axial, and also in the case of the α-anomer of a sugar, it is clear that the destabilizing C–O dipole effects, present when oxygen of OMe is equatorial, are minimized when this oxygen becomes axial. The absence of any significant dipolar repulsions in the latter case has been proposed as the cause of the anomeric effect (see Perrin et al.[7]). Though its cause is still uncertain the anomeric effect is well documented and it commonly has a magnitude of ca. 8–10 kJ mol^{-1}, and is solvent dependent.

Worked Problems

Q1. Give stereochemical diagrams for: (i) cis-cyclobutane-1,2-dicarboxylic acid (**56**); (ii) trans-cyclobutane-1,2-dicarboxylic acid (**57**); (iii) cis-cyclobutane-1,3-dicarboxylic acid (**58**); (iv) trans-cyclobutane-1,3-dicarboxylic acid (**59**). Only one of the above compounds is chiral. Give the R/S configuration of the stereoisomer that you draw for the only chiral molecule in (i)–(iv).

A. See structures below. The dihedral angles between C(1)C(4)C(3) and C(3)C(2)C(1) in **56**, and corresponding angles in **57** and **58**, are 148–156°. The cyclobutane ring in **59** is planar. The only chiral molecule is **57**; the configuration shown is (1S,2S).

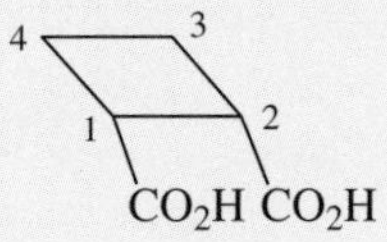

56 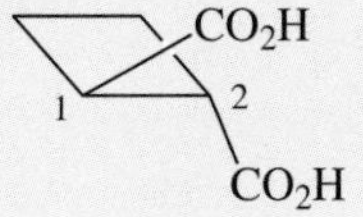   **57**

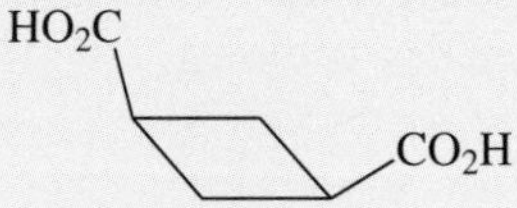

58 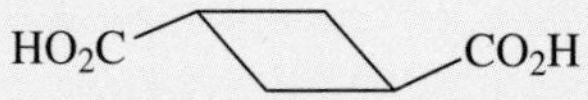   **59**

Q2. With the aid of molecular models, decide which of the following you would expect to be chiral: (i) *trans*-cyclopentane-1,2-dicarboxylic acid; (ii) *cis*-cyclopentane-1,2-dicarboxylic acid; (iii) *trans*-cyclopentane-1,3-dicarboxylic acid; (iv) *cis*-cyclopentane-1,3-dicarboxylic acid.

A.   *trans*-Cyclopentane-1,2-dicarboxylic acid (**60**) and *trans*-cyclopentane-1,3-dicarboxylic acid (**61**) are chiral. The *cis*-isomers are *meso*.

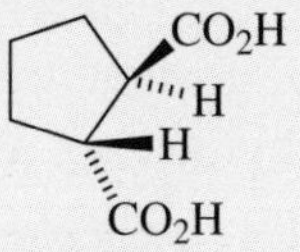

60 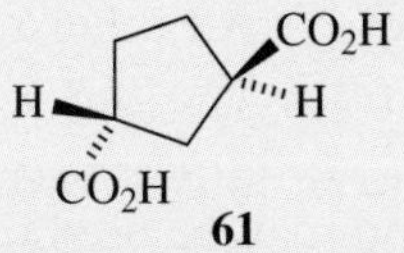   **61**

Q3. Show each of the following compounds in its most stable chair conformation: (i) *trans*-2-fluoro-1-isopropylcyclohexane (**62**); (ii) *cis*-1-*t*-butyl-3-fluorocyclohexane (**63**); (iii) *cis*-1-chloro-4-methylcyclohexane (**64**); (iv) *trans*-4-fluorocyclohexanecarboxylic acid (**65**); (v) 1,1-dichloro-4-fluorocyclohexane (**66**); (vi) *trans*-4-*t*-butylcyclohexanol (**67**); (vii) *cis*-cyclohexane-1,2-dicarboxylic acid (**68**); (viii) *trans*-3-chlorocyclohexanol (**69**).

A. See structures below.

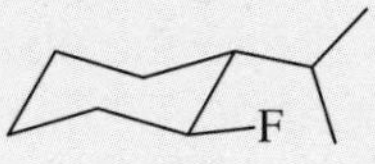

62 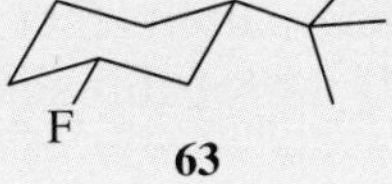   **63** 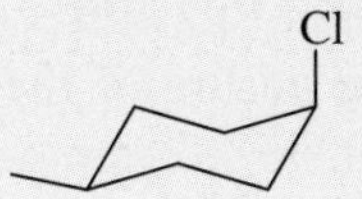   **64**

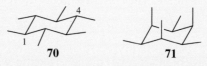

65   66   67

68   69

Q4. Give names to **70** and **71**, and with the aid of molecular models state whether either of these molecules is chiral.

70   71

A. Compound **70** can be named in three different ways. Consider the methyl groups; each is *trans* to both neighbours and **70** is often referred to as all-*trans*-hexamethylcyclohexane. It can also be called 1α,2β,3α,4β,5α,6β-hexamethylcyclohexane. This use of α (down) and β (up) is comparable to the terminology of steroids and the choice of prefix α or β is based on the orientation with respect to a plane through the cyclohexane ring. This nomenclature is preferred by *Chemical Abstracts*. Thirdly, **70** can be named by taking one methyl group, say that attached to C(1), as a reference and giving it the symbol '*r*'. All other methyl groups are assigned stereochemistry with respect to the '*r*' methyl group. With this method **70** is *r*-1,*trans*-2,*cis*-3,*trans*-4,*cis*-5,*trans*-6-hexamethylcyclohexane. The same methods can be used to denote the stereochemistry in **71**, which is (1) all-*cis*-hexamethylcyclohexane or (2) 1α,2α,3α,4α, 5α,6α-hexamethylcyclohexane. That this assignment is correct can be seen with a molecular model of **71**; this shows that all methyl groups are inclined towards the plane that passes through the three axial hydrogens of a notional cyclohexane ring; also, for a fresh perspective, turn the model through 90° in a clockwise (or anticlockwise) sense. Compounds **70** and **71** are not chiral. This is because a plane of symmetry passes through C(1) and C(4), and similarly through C(2) and C(5), and C(3) and C(6) of both molecules. Compound **71** represents an interesting case, as it can be thought of as three repeating pairs of *cis*-1,2-dimethyl units, and it is the nature of the repeat pattern that gives rise to the planes of symmetry. Compound **71** is probably simpler than *cis*-1,2-dimethylcyclohexane, in which the mirror image of a chair conformer appears not to be superimposable on the original, and only becomes identical after ring inversion.

Q5. Draw the more stable chair conformation of **72** and name the compound. Is this compound chiral?

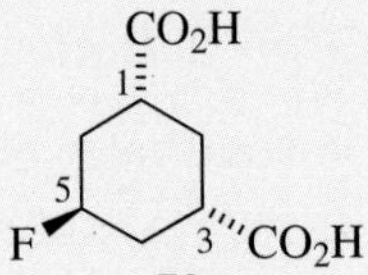

A. Compound **72** is drawn in **73** in its most stable chair conformation, with the bulkier CO_2H groups equatorial and the smaller F axial. The name is *trans*-5-fluorocyclohexane-*r*-1,*cis*-3-dicarboxylic acid or 5β-fluorocyclohexane-1α,3α-dicarboxylic acid. A plane of symmetry through C(2) and C(5) of **72** rules out chirality.

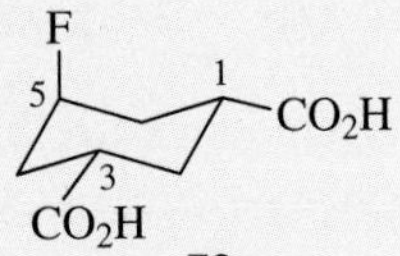

Q6. Draw the chair conformation of *r*-1-*cis*-4-di-*t*-butyl-*cis*-2-*cis*-5-dihydroxycyclohexane. With the aid of a model, consider whether the molecule would exist entirely in the chair conformation.

A. The chair conformation is shown in **74**. One consequence of a chair conformation is that a pronounced 1,3-diaxial interaction exists between a *t*-butyl group and a hydroxyl group. This can be avoided if the ring adopts a twist-boat conformation **75**. The number of molecules that exist to any extent in the twist-boat conformation is small, and **75** has the benefit of a hydrogen-bonded interaction between the two hydroxyl groups. The twist-boat conformation is *ca.* 6 kJ mol^{-1} more stable than the boat conformation. In this work, Stolow[8] claims an equilibrium between **74** and **75**. Another molecule known to exist in a twist-boat conformation is cyclohexane-1,4-dione (**76**) (see Hoffmann and Hursthouse[9]).

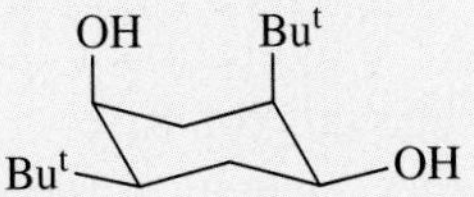

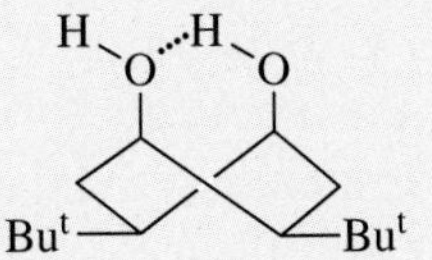

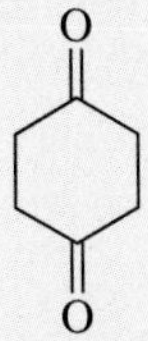

Q7. *trans*-3-Methylcyclohexanol can exist in two chair conformations, **28** and **29**. Use the conformational free energies (*A* values) of Me (6.70 kJ mol^{-1}) and OH (3.75 kJ mol^{-1}) and calculate which is the more stable chair conformer, and what are the relative populations of **28** and **29**.

A. The A value for Me is larger than that for OH and so **28** will be the more abundant conformer. On the basis of the A values one can state that:

$$\Delta G^{\circ}_{calc} = -3.77 - (-6.70) = 2.93 \text{ kJ mol}^{-1}$$

This is in fair agreement with the experimental value of 5.02 kJ mol^{-1}.

Summary of Key Points

- In this chapter the stereochemistry of disubstituted cyclopropanes, cyclobutanes, cyclopentanes and cyclohexanes was considered.
- With two identical substituents in the *cis*-1,2 positions, cyclopropanes, cyclobutanes, cyclopentanes and cyclohexanes are *meso*; when the substituents are *trans* the corresponding molecules are chiral.
- Cyclobutane is not flat. *Cis*- and *trans*-1,3-disubstituted cyclobutanes are diastereoisomers, though neither is chiral. This unusual situation applies also to 1,4-disubstituted cyclohexanes. When the substituents are identical, *cis*-1,3-disubstituted cyclopentanes and *cis*-1,3-disubstituted cyclohexanes are *meso*; the corresponding *trans* isomers are chiral.
- Cyclopentane exists in a puckered conformation, reminiscent of a square envelope; the puckering rotates rapidly around the ring.
- When the substituents X and Y in a disubstituted cyclohexane are 1,2-diaxial (dihedral angle X–C–C–Y is 180°) the configuration is *trans*. The same holds for the 1,4-diaxial case, but now we measure the dihedral angle with an imaginary C(1)···C(4) bond. When X and Y are 1,3-diaxial, the dihedral angle, using an imaginary C(1)···C(3) bond, is 0° and the configuration is *cis*. When X and Y are axial/equatorial the configuration is the opposite of the corresponding diaxial or diequatorial case.
- In more extensively substituted cyclohexanes, one substituent is taken as a reference and given the symbol '*r*'. Other substituents are defined as *cis* or *trans* with respect to this reference.
- *Trans*- and *cis*-decalin differ in the stereochemistry of the ring junction. *trans*-Decalin has the ring junction hydrogens *trans*

diaxial; in *cis*-decalin the ring junction hydrogens are axial and equatorial with respect to one six-membered ring and equatorial and axial with respect to the other. Whereas *trans*-decalin is rigid about the ring junction, *cis*-decalin can undergo inversion by exploiting the flexibility around this central bond.

* Steroids have a *trans* A/B ring junction in the majority of cases; in bile acids, however, this junction is *cis*.

* In some sugars, *e.g.* D-glucose, and 2-alkoxytetrahydropyrans, the axial form is populated to a greater extent than the axial population of the corresponding cyclohexane would suggest. This is known as the anomeric effect and two possible contributory factors were discussed.

Problems

6.1. With the aid of molecular models and diagrams, indicate whether you would expect the following to be chiral: (a) *trans*-cyclohexane-1,2-dicarboxylic acid; (b) 3-methylcyclohexanone; (c) 1,1,4-trichlorocyclohexane; (d) 1,1,2-trichlorocyclohexane; (e) *cis*-3-methylcyclohexanecarboxylic acid; (f) 1,1,3,3-tetramethylcyclohexane; (g) 1,1,3-trimethylcyclohexane; (h) the anti-diabetic compound 1,2,3-cyclohexanetrione.

6.2. What is the stereochemical relationship between the carboxylic acid groups in **77**? Would it be possible to make an anhydride from **77**?

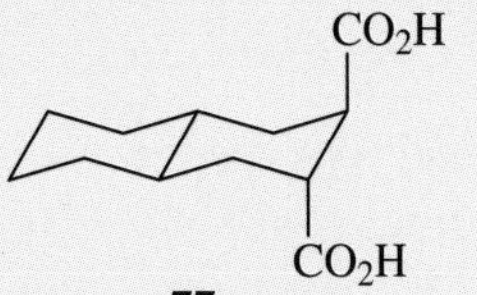

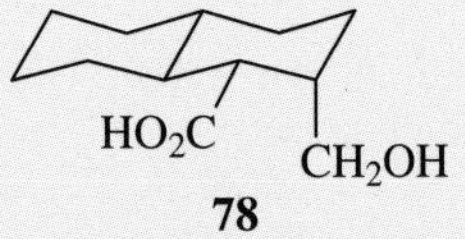

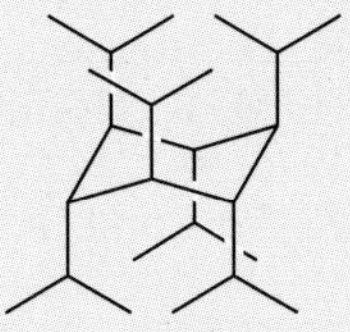

6.3. Draw *trans*-1,3-difluorocyclohexane. Is this compound chiral?

6.4. What is the stereochemical relationship between the carboxylic acid and hydroxymethyl (CH_2OH) groups in **78**? Is it possible to form a lactone from this hydroxy acid?

6.5. The central structures of all-*trans*-hexamethylcyclohexane (**70**) and the hexaethyl analogue show that all the alkyl groups exist in the equatorial position. However, the crystal structure of the all-*trans*-hexaisopropylcyclohexane (**79**) indicates that every isopropyl group is axial. With the aid of space-filling molecular models, examine both the all-equatorial and all-axial structures and assess whether the unexpected experimental finding is reasonable (see Gore and Biali[10]).

References

1. W. A. Bernett, *J. Chem. Educ.*, 1967, **44**, 17.
2. P. von R. Schleyer, J. E. Williams and K. R. Blanchard, *J. Am. Chem. Soc.*, 1970, **92**, 2377. In this paper the strain in cyclohexane is given as 5.6 kJ mol^{-1}, and that in *trans*-decalin as 7.5 kJ mol^{-1}.
3. D. H. R. Barton, *Topics Stereochem.*, 1971, **6**, 1. This contains a reprint of Barton's original publication in *Experientia* (1950) in which the term 'conformational analysis' was first used. Initially 'equatorial' and 'polar' were used to describe the positions of exocyclic hydrogens, considered as substituents. Subsequently, the term 'polar' was replaced by 'axial' at the suggestion of Ingold. This is described in a very readable account of many aspects of stereochemistry (including a brief discourse on the origin of the CIP convention) by V. Prelog in '*My 132 Semesters of Chemistry Studies*', American Chemical Society, Washington, 1991, p. 78.
4. S. Winstein and N. J. Holness, *J. Am. Chem. Soc.*, 1955, **77**, 5562.
5. H. Senderowitz, C. Parish and W. C. Still, *J. Am. Chem. Soc.*, 1996, **118**, 2078.
6. A. J. Kirby, *The Anomeric Effect and Related Stereo-electronic Effects at Oxygen*, Springer, Berlin, 1983, p. 75.
7. C. L. Perrin, K. B. Armstrong and M. A. Fabian, *J. Am. Chem. Soc.*, 1994, **116**, 715 (anomeric effect).
8. R. D. Stolow, *J. Am. Chem. Soc.*, 1961, **83**, 2592.
9. H. M. R. Hoffmann and M. B. Hursthouse, *J. Am. Chem. Soc.*, 1976, **98**, 7449.
10. Z. Gore and S. E. Biali, *J. Am. Chem. Soc.*, 1990, **112**, 893.

Further Reading

V. G. S. Box, *Heterocycles*, 1990, **31**, 1157.
J. D. Dunitz, *Tetrahedron*, 1972, **28**, 5459.
J. D. Dunitz and J. Weser, *J. Am. Chem. Soc.*, 1972, **94**, 5645. These papers describe conformational parameters and geometric constraints in five and six membered rings.
E. L. Eliel and S. H. Wilen, *Stereochemistry of Organic Compounds*, Wiley, New York, 1994, pp. 665 and 686.
B. Fuchs, *Topics Stereochem.*, 1978, **10**, 1.

7

Substitution Reactions at Saturated Carbon

Aims

By the end of this chapter you should be aware:

- That the S_N2 reaction always involves inversion of configuration
- That S_N2 reactions always occur for primary substrates (RCH_2X), are very uncommon for tertiary compounds, and are well documented for secondary compounds, depending on the nucleophile, substrate and solvent
- That the S_N2 reaction will not occur at bridgehead positions
- That intramolecular nucleophilic displacements occur readily if they are 'exocyclic', but there are limitations on 'endocyclic' counterparts
- That the S_N2' reactions of allylic (propenyl) systems occur with variable stereochemistry
- That product from an S_N1 reaction can exhibit a range of stereochemistries from complete retention to partial inversion

7.1 Nucleophilic Substitution

7.1.1 The S_N2 Reaction

In this type of reaction a group X at a saturated carbon is replaced by a group Y. Typically, X = Cl, Br, I, $^+OH_2$, $OSO_2C_6H_4Me$-4 (OTs), $OSO_2C_6H_4Br$-4 (OBs) or $^+SR_2$; examples of good nucleophiles, Y, include HO^-, RO^- (R = alkyl), Me_3N, ^-CN, RS^- or I^-. Note that iodide is both a good leaving group and a good nucleophile. This heterolytic reaction (which involves two-electron shifts) occurs by one of two mechanisms.

These are distinguishable by both kinetic and stereochemical criteria; the kinetic criteria are employed for the designation of the reactions as S_N1 (substitution, nucleophilic, unimolecular) and S_N2 (substitution, nucleophilic, bimolecular).

Firstly, we consider the S_N2 reaction at saturated carbon, which is a one-step process that always proceeds with inversion of configuration. This reaction is shown in Scheme 7.1 with the inversion taking place at C(2) of compound **1**. Inversion of configuration in what we now know as S_N2 reactions was first suspected in 1895 by the Latvian-born chemist Walden. He reacted hydroxysuccinic acid (**2**, malic acid) with the acyl chlorides shown in Scheme 7.2, and realized that *one* of the steps shown in Scheme 7.2 was associated with inversion of configuration, but at that time he was unable to specify which.

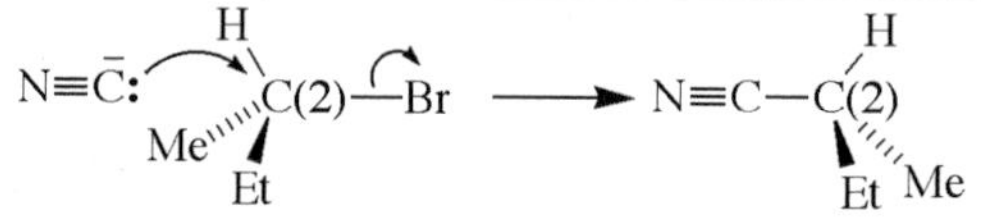

Scheme 7.1

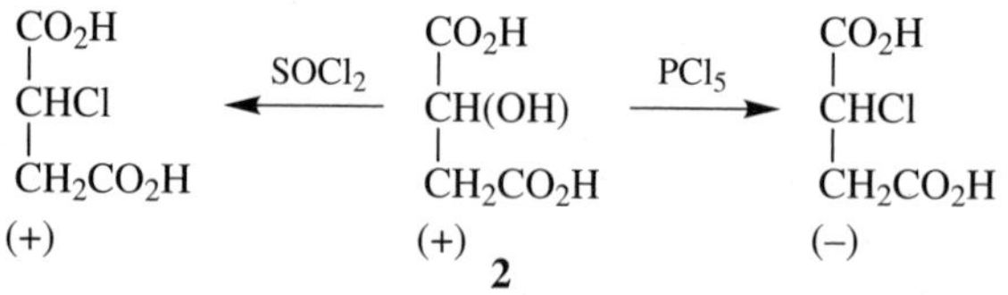

Scheme 7.2

Proof that an S_N2 reaction proceeds with inversion of configuration was provided by Kenyon *et al.*[1] in a three-step reaction sequence in which, importantly, the starting material and the product were enantiomers. The reaction sequence is shown in Scheme 7.3, and has the following features: (1) compound **3** has a stereogenic centre; (2) it is converted into the 4-methylbenzenesulfonate (also known as toluene-4-sulfonate or tosylate, and abbreviated in structures as OTs) ester **4** by the appropriate acyl chloride in a reaction that does not involve bond cleavage at the stereogenic centre; one can therefore say that **4** and **3** have the same configuration; (3) transformation of **4** into the ethanoate (acetate) **5** involves displacement of OTs by ethanoate ion, and it is here that inversion of configuration takes place. Note that ethanoate ion is a good nucleophile, in contrast to the poor nucleophilicity of ethanoic (acetic) acid; (4) the final step, hydrolysis of the ethanoate to produce **6**, occurs with retention of configuration, and this is now known to be the case under the conditions used for the hydrolysis of most ethanoates.

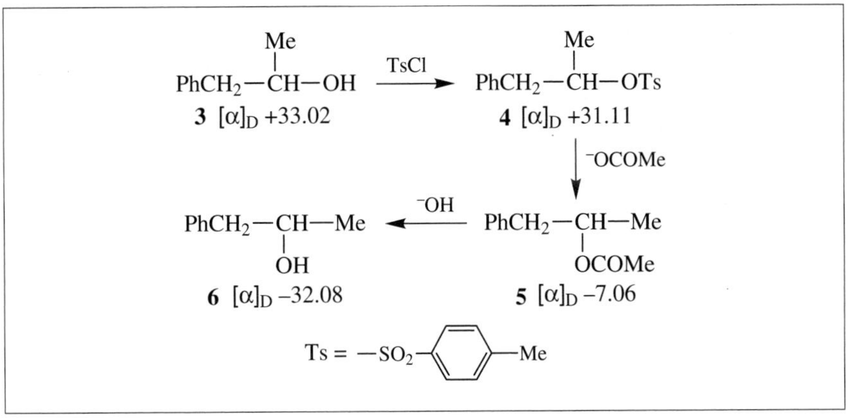

$$
\text{PhCH}_2\text{—CH—OH} \quad \xrightarrow{\text{TsCl}} \quad \text{PhCH}_2\text{—CH—OTs}
$$
$$
\overset{\text{Me}}{|} \qquad\qquad\qquad \overset{\text{Me}}{|}
$$

3 $[\alpha]_D$ +33.02   **4** $[\alpha]_D$ +31.11

$\downarrow$ $^-$OCOMe

$$
\text{PhCH}_2\text{—CH—Me} \quad \xleftarrow{\ ^-\text{OH}\ } \quad \text{PhCH}_2\text{—CH—Me}
$$
$$
\overset{}{\underset{\text{OH}}{|}} \qquad\qquad\qquad \overset{}{\underset{\text{OCOMe}}{|}}
$$

6 $[\alpha]_D$ –32.08   **5** $[\alpha]_D$ –7.06

$$
\text{Ts} = -\text{SO}_2\!-\!\!\left\langle\ \bigcirc\ \right\rangle\!\!-\text{Me}
$$

Scheme 7.3

The above example, a so-called Kenyon–Phillips cycle, was one of a number that this group carried out, including some with the more powerful nucleophile EtO⁻, and self-consistent results were obtained throughout. Apart from their role in providing proof that concerted nucleophilic reactions proceed with inversion of configuration, analogues of the Kenyon–Phillips cycles have been very useful in inverting the configuration of alcohols.

After this work, a further elegant experiment was carried out by Hughes et al.,[2] with the measurement of the second-order rate constant for a concerted nucleophilic substitution reaction, and this was done in two ways. The substrate was one enantiomer of 2-iodooctane (**7**) and the nucleophile was radioactive iodide anion, *I⁻, in propanone (acetone). The nature of the reaction is outlined in Scheme 7.4. Firstly, the rate constant was determined polarimetrically to give a rate constant k_α; then the rate constant for exchange of *I⁻ was determined; this is represented by k_{ex}. The ratio of these rate constants within experimental error was 2:1.

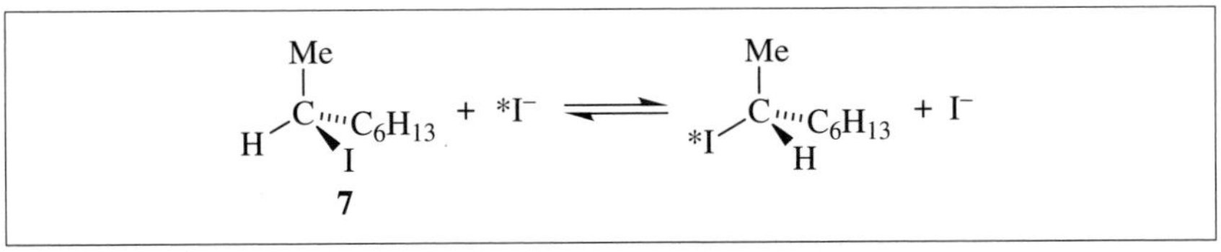

$$
\underset{\text{H}}{\overset{\text{Me}}{\text{C}}}\!\!\cdots\text{C}_6\text{H}_{13} \;+\; {}^*\text{I}^- \;\rightleftharpoons\; {}^*\text{I}\!-\!\!\underset{\text{H}}{\overset{\text{Me}}{\text{C}}}\!\!\cdots\text{C}_6\text{H}_{13} \;+\; \text{I}^-
$$

7

Scheme 7.4

Now each act of nucleophilic substitution (1) occurs at a rate that is in accord with the observed value of k_{ex} and (2) produces a molecule of inverted configuration, and the polarimeter monitors this aspect of the reaction. For, say, the first 1% of reaction, the polarimeter will indicate 98% of the original optical activity, as the solution now contains 2% of the racemic compound. This explains why the polarimetric rate constant is twice that for exchange of the radiolabel (see Hughes et al.[2]).

The S_N2 reaction at saturated carbon always proceeds with inversion of configuration.

Inversion of configuration always accompanies S_N2 reactions; the transition state is shown schematically in Figure 7.1. An everyday analogy of this inversion is shown by an umbrella that blows inside out on a windy day; of course, the umbrella has more spokes than a carbon atom has valencies. This analogy has some further merit in that the nucleophile, the 'central' carbon and the leaving group are collinear, and remain so in the transition state for the reaction.

The S_N2 reaction is almost always observed for primary halogenoalkanes (alkyl halides) **8**, tosylates **9** and other related compounds. Inversion of stereochemistry in the S_N2 reaction of primary compounds has been investigated by the use of deuterium to create a stereogenic centre in, for example, **10**.

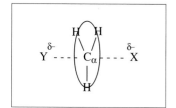

Figure 7.1 Transition state for the S_N2 reaction $Y^- + Me-X \rightarrow$ $Y-Me + X^-$

$$RCH_2-X$$

8 X = hal

9 X =OSO_2—⟨benzene ring⟩—Me

$$Me-C\overset{H}{\underset{X}{\big|}}D$$

10

Most secondary compounds, such as **11** and **12**, will undergo the S_N2 reaction (in which the substituents are alkyl groups). When at least one of the groups is aromatic there is some tendency for the group X, the leaving group, to leave of its own accord (this is the basis of the S_N1 reaction, Section 7.1.3) rather than to be expelled by the nucleophile in a concerted S_N2 reaction. However, the experiments that defined the stereochemistry of the S_N2 reaction were performed on secondary substrates.

For tertiary substrates, *e.g.* *t*-butyl chloride (**13**, 1-chloro-1,1-dimethylethane), significant non-bonded steric repulsion prevents an incoming nucleophile from approaching carbon C(1) from the rear to form the collinear arrangement of atoms required in the S_N2 reaction. For this reason, tertiary substrates instead undergo the alternative S_N1 pathway for substitution.

This question of collinearity is paramount for S_N2 reactions. There are even some primary substrates for which the S_N2 reaction occurs very slowly on account of severe non-bonded interactions experienced by the nucleophile as it attempts to gain access to the carbon that is bonded to the leaving group while maintaining the necessary collinearity shown in Figure 7.1. One important example is neopentyl chloride (**14**, 1-chloro-2,2-dimethylpropane).

$$Me\diagdown_{Et}C\diagup^H_X$$

11 X = hal

12 X =OSO_2—⟨benzene ring⟩—Me

In S_N2 reactions the nucelophile, the carbon being attacked and the leaving group must be collinear.

⟨structure⟩—Cl

13

⟨structure⟩—CH_2Cl

14

S_N2 reactions at some primary carbons can be very slow if it is difficult for the nucleophile to approach the carbon carrying the leaving group.

Make a model of compound **14** and satisfy yourself that establishment of collinearity between a nucleophile, *e.g.* HO⁻ or Me₃N, is rather difficult.

Table 7.1 Relative rate constants for the reaction R–I + Cl⁻ → R–Cl + I⁻ in propanone

R	Rate
Me	1.00
Et	0.089
Bu^i	0.0034
CH$_2$Bu^t	1.2×10^{-6}

15

Table 7.1 shows relative rate constants for halide exchange reactions in acetone, and the progressively slower reactions observed as one looks down this table are purely a reflection of the difficulty in achieving the requisite collinear three-atom unit (*cf.* Figure 7.1).

Caged compounds in which the leaving group is located at a bridge-head position, *e.g.* **15**, do not undergo S$_N$2 reactions for two reasons: (1) the cage structure prevents approach of the nucleophile to C(1) by the required route; (2) even if it could, it is not possible to invert the configuration of the carbon that is bonded to the leaving group.

In all the cases covered so far, the nucleophile acts as an entity that is separate from the leaving group X, and the carbon, C$_\alpha$, that the nucleophile attacks. Can the nucleophile be part of the same molecule as C$_\alpha$ and X? Such a reaction would, of course, be a concerted nucleophilic displacement, but since both the nucleophile and the leaving group are part of the same molecule, the term S$_N$2 is not in order. In such intramolecular displacements, two sub-categories can be envisaged. Eschenmoser's group describe these in a paper (see Tenud *et al.*[3]), and the schematic representations are termed exocyclic (Figure 7.2a) and endocyclic (Figure 7.2b), respectively. In Figure 7.2, N and L refer to the nucleophile and leaving group, respectively, and C$_\alpha$ is the carbon that is subject to nucleophilic attack.

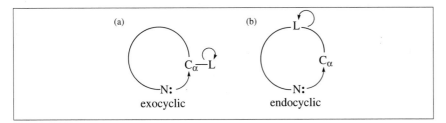

Figure 7.2

There is little restriction on intramolecular exocyclic substitutions; this is because N, C$_\alpha$ and L can readily become collinear. This does not hold, however, in the case of the endocyclic reaction. An important question is: what is the smallest size of ring in the transition state for a concerted endocyclic substitution?

Eschenmoser's group addressed this question with the carbanion **16** and its hexadeuterio counterpart **17**, in which two separate methyl groups are labelled as CD_3 and the carbanion brings about a concerted nucleophilic displacement, endocyclic or exocyclic, at a sulfonate methyl group.

> Why is it necessary to have two groups labelled in this experiment?

An equimolar mixture of **16** and **17** was allowed to react and the carbanion acts as a nucleophile; the sulfonate ester acts as the leaving group. The product can be described as D_0, D_3 or D_6 according to the number of deuterium atoms that it contains. If the reaction proceeds by an endocyclic substitution, which is intramolecular, the deuterium labelled groups will not be separated and the product distribution will be D_0 and D_6 in equal amounts, *i.e.* D_0:D_3:D_6 is 1:0:1.

An intermolecular reaction of the above equimolar mixture will give mainly D_3, but also some D_0 and D_6 since in this case the ion reacts with either an unlabelled molecule (no deuterium), or a labelled molecule (six deuterium atoms). The distribution from an intermolecular reaction is given by the ratio D_0:D_3:D_6 = 1:2:1. The two mechanisms can readily be distinguished by mass spectrometric analysis of the product. The result indicated that the reaction was *intermolecular*. The reason for this is that in the transition state for substitution it is not possible to have a collinear arrangement for the atoms, C–C–O, in **16** that corresponds to N, C_α and L in Figure 7.2b.

This raises the question: what is the smallest ring size that can be tolerated in the transition state for the intramolecular endocyclic nucleophilic substitution as shown in Figure 7.2b? The answer was shown by two experiments.

Firstly, an alkaline solution of the thebaine derivative **18** is rapidly converted into **19** by a reaction that involves an eight-membered endocyclic transition state, if one counts around the smallest available ring. Secondly, in a systematic investigation, low concentrations of the amino ester **20** were converted into **21** to the extent of 16% by an intramolecular nine-membered endocyclic pathway. The lower homologue **22**, however, undergoes a corresponding reaction exclusively by way of an

Exocyclic concerted nucleophilic displacements (Figure 7.2a) occur readily. The endocyclic counterparts are uncommon, and the smallest cyclic transition state for this type of substitution is eight membered.

intermolecular reaction. It is interesting that whereas the endocyclic eight-membered transition state cannot be formed from **20**, it is not a problem in the reaction of **18**, presumably because here the constraining multiple ring system assists in bringing about the necessary collinear alignment, corresponding to N, C_α and L in Figure 7.2b (see King and McGarrity[4] and Kirby *et al.*[5]).

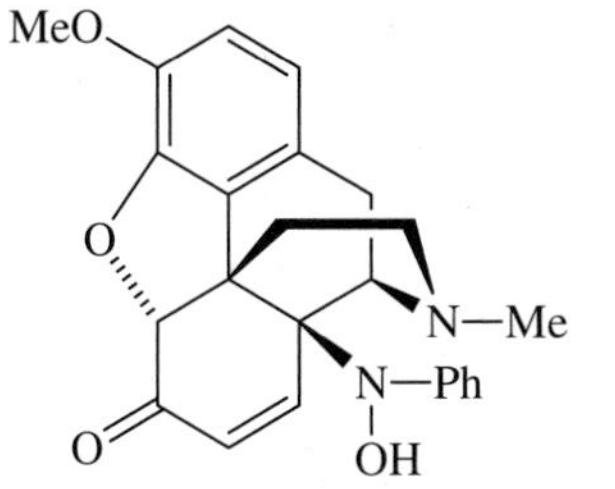

18

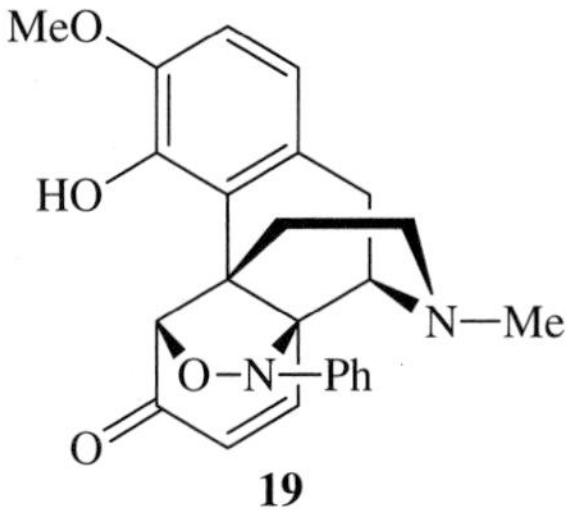

19

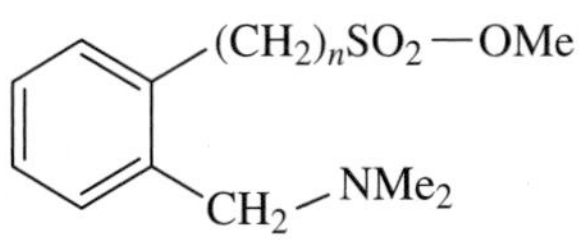

20 $n = 2$
22 $n = 1$

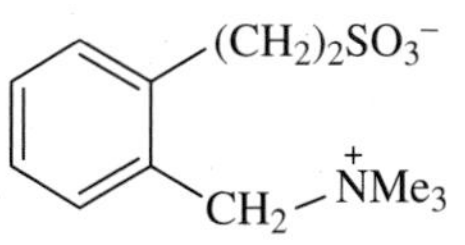

21

7.1.2 The S$_N$2′ Reaction

When an allylic compound such as **23** is subjected to nucleophilic attack as shown in Scheme 7.5, the product is that of the S$_N$2 reaction. If the allylic compound now has the structure shown in **24**, there is significant steric hindrance to approach of the nucleophile to the chlorine-bearing carbon. There is, however, an alternative site, available to the nucleophile, and this is shown in Scheme 7.6.

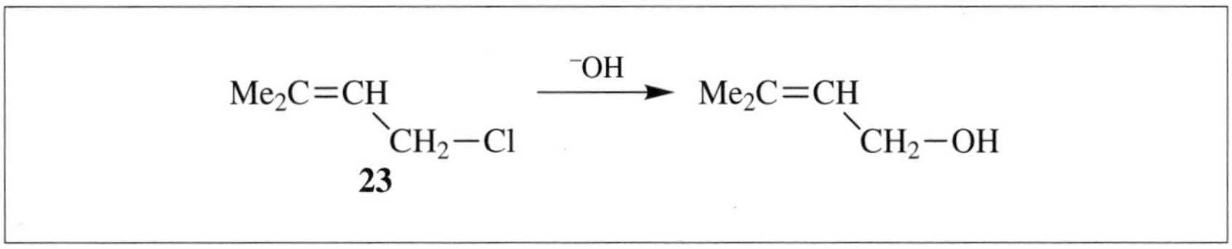

Scheme 7.5

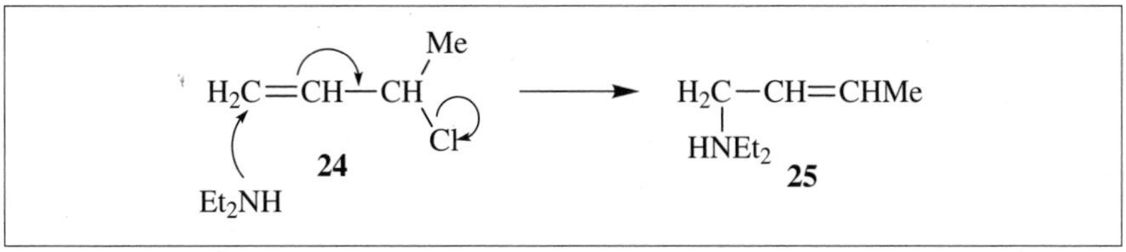

Scheme 7.6

This involves attack by a nucleophile at the terminal carbon of the double bond with the leaving group, chloride, expelled by an intramolecular relay of electrons along the molecule. As a result, the product is **25**. A similar series of electron shifts is involved in the initial step of the Michael addition.

The stereochemistry of the S_N2' reaction has been shown to be variously *syn* (the nucleophile and leaving group are on the same face of the allylic system) or *anti*, depending on the nature of the nucleophile and leaving group. For example, the *cis* isomer **26** with piperidine gave **27**, in a clean *syn* allylic displacement, whereas Na^+SPr gave a *ca.* 2:1 mixture of isomers in which the *trans* isomer **28** was the major product. This trend toward *anti* substitution with RS^- is more pronounced (almost exclusive) in the acyclic case **29**.

S_N2' reactions occur in allylic systems when the direct S_N2 reaction is sterically inhibited. There are no definitive rules of stereochemistry, though with RS^- *anti* stereochemistry is pronounced.

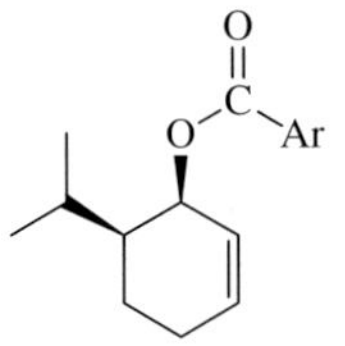

26
(Ar = aromatic group)

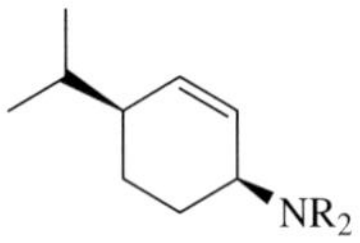

27

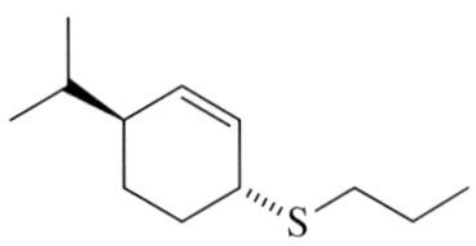

28

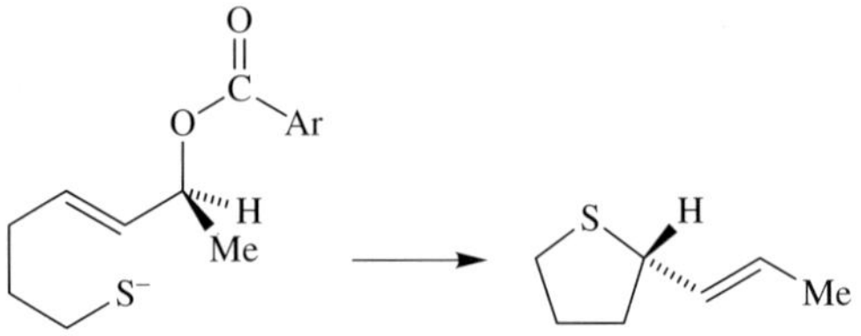

29

30

Use molecular models to show that *anti* substitution of **29** leads to **30**. It may be helpful to make a model of **30** first, in order to see more clearly how '*anti*' attack on **29** initiates a series of bond changes that result in **30**.

7.1.3 The S_N1 Reaction

In contrast to the S_N2 reaction, which always shows clear-cut inversion of configuration at the carbon that is attacked by the nucleophile, the stereochemical outcome of the S_N1 reaction is variable.

In this second type of nucleophilic substitution at saturated carbon, a two-stage reaction occurs in which the leaving group departs to give a carbocation (that is, two electrons short of an octet) as an intermediate.

The S_N1 reaction shown in Scheme 7.7 consists of a slow initial step followed by a second and rapid step in which the carbocation regains its octet of electrons by reaction with a nucleophile.

The S_N1 reaction is almost always followed by tertiary halides, and also by secondary halides and, for example, tosylates, especially when at least one of the substituents is an aryl group. From this latter case the carbocation is now benzylic. The nature of the intermediate carbocation in an S_N1 reaction is illustrated in Figure 7.3, which shows the three substituents in a coplanar arrangement and with an empty p-orbital. This structure provides a clue as to the stereochemistry of the subsequently formed tetrahedral product.

If the carbocation has a long life, one would expect a racemic product. The nucleophile, which may include solvent, *e.g.* ethanol, can approach with equal probability from either face. One can describe the carbocation as prochiral (see Chapter 8), though it is not easy to realize large preferences for one enantiomer over the other in the product. However, it has been known for some time that the product is not always fully racemic, and an example is the ethanoate (acetate) ester derived from reaction of one enantiomer of 1-phenylethyl tosylate (**31**) with ethanoic (acetic) acid; such reactions are known as acetolyses. The product **33** is formed with *ca.* 12% net inversion. Since ethanoic acid ($MeCO_2H$) is a poor nucleophile (unlike the ethanoate anion, $MeCO_2^-$), this stereochemical result is most likely to arise from shielding of one face of the carbocation as it develops, by the anion ^-OTs. The shielding means that the incoming solvent, acting as a nucleophile, will attack the carbocation to give product with the net inversion stated.

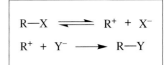

Scheme 7.7

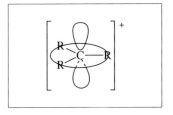

Figure 7.3 Diagram of a carbocation with three R groups coplanar and a vacant p-orbital

S_N1 reactions are always observed for tertiary substrates; in open-chain systems, excess inversion is often observed because the leaving group shields the planar carbocation from attack on one face.

Ph   H
  \\C/
Me/ \\X

31 X = OTs
32 X = Cl

Ph   OCOMe
  \\C/
Me/ \\H

33

Bu^t Ph
   \\C/
 H/ \\X

34 X = OTs
35 X = OAc

In a slightly later paper, Steigman and Hammett[6] found a corresponding figure of 15% net inversion for the product of acetolysis of the corresponding chloride **32**. Winstein and Morse[7] reported a value of 10% net inversion for the ethanoate **35**, obtained after reaction of **34** with ethanoic acid.

7.1.4 S_N1 Reactions with Retention of Configuration

The preferred structure of a carbocation is planar, and the nature of the bridgehead position is such that any carbocation formed at the bridgehead is unable to become planar. This raises the question of how easy is it to form a bridgehead carbocation? The role of kinetics is significant

36 X = OTs
37 X = H

The S$_N$1 reactions of molecules with leaving groups at the bridgehead are slow because the carbocation *cannot* become planar.

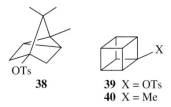

OTs
38

39 X = OTs
40 X = Me

here. Consider the bridgehead adamantyl tosylate (**36**). The parent hydrocarbon adamantane, C$_{10}$H$_{16}$, can be constructed with models from the all-axial conformation of *cis,cis*-1,3,5-trimethylcyclohexane; if one bonds the three methyl groups of this molecule to a single carbon, adamantane (**37**) results. As inspection of a model reveals, there are four fused chair cyclohexane rings in adamantane. Adamantane is the basis of the diamond structure and has an extremely high melting point, 268 °C, for its relative molecular mass.

In the S$_N$1 reaction of **36**, departure of the tosylate group leaves a carbocation that cannot become planar. The molecule **36** responds to this situation by forming the carbocation more slowly than might be expected for an analogous acyclic molecule. However, the S$_N$1 reaction does occur; this is in contrast to the corresponding S$_N$2 reactions at bridgehead positions which are totally ruled out. The slowest S$_N$1 reaction from a tertiary bridgehead substrate is exhibited by the tricyclyl structure **38**, whose rate constant for ionization is *ca.* 10^{-14} that of **36**. The rate retardation is almost solely due to the bond angles at C(4) of **38** being more 'tied-back' than in **36**; that is, they are further from the 120° that is preferred for carbocations (see Sherrod *et al.*[8]).

Interestingly, there is one substrate that is predicted by molecular mechanics calculations to ionize still more slowly than **38**. This compound is **39**, a derivative of the hydrocarbon known as cubane, and in which the CCC bond angles are 90°. However, in contrast to **36**, and especially **38**, the ester **39** undergoes S$_N$1 reaction in methanol much faster than expected. The product from **39** is the corresponding methyl ether **40** (see Eaton[9]).

One of the reasons that tosylates, or related sulfonate esters, are employed as leaving groups rather than, say, chlorides is that from an enantiomerically pure alcohol, with a stereogenic carbon that is bonded to OH, the tosylate is also enantiomerically pure since the bond between oxygen and the stereogenic carbon is not broken.

Chloride is a convenient leaving group, and alkyl chlorides (chloroalkanes) are usually prepared from an alcohol, by reaction with thionyl chloride (sulfur dichloride oxide), SOCl$_2$. The reaction is outlined in Scheme 7.8; step (1) involves formation of a chlorosulfite ester **41**; this contains a good leaving group, –O–S(O)–Cl, which is displaced with inversion of configuration to give **42** and a chloride ion which then acts as a nucleophile toward **42**; the alkyl chloride (chloroalkane) **43** is formed, again with inversion of configuration. Therefore the overall formation of **42** from ROH involves two inversions of configuration, and this corresponds to net retention. Butan-2-ol is known to be converted into 2-chlorobutane with *ca.* 98% retention of configuration by thionyl chloride in dioxane.

The bicyclic brosylate (4-bromobenzenesulfonate) **44** undergoes acetolysis to give ethanoate **46**, with complete retention of configuration.

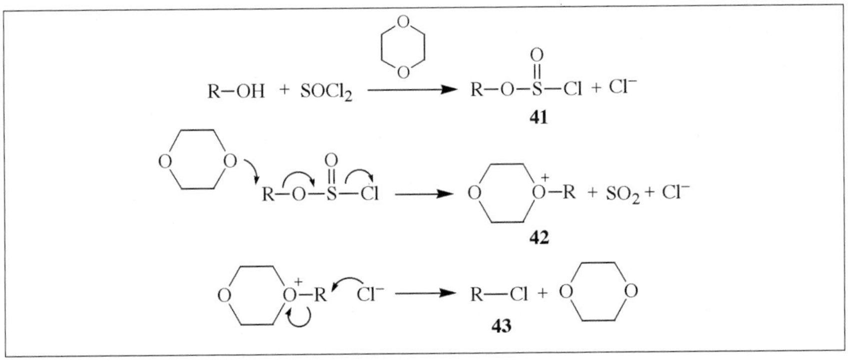

Scheme 7.8

This sequence of events is explained as follows. The leaving group is expelled with assistance of the suitably oriented π-bond present in the same molecule, as indicated in Scheme 7.9; a delocalized carbocation **45** is formed. Note that the electrons of the π-bond of **44** act as an internal nucleophile, to produce **45** in a concerted reaction with inversion of configuration. The reaction of **45** to give **46** can again be regarded as a concerted reaction with inversion of configuration, in which two partial bonds to C(7) in **45** act as a leaving group with respect to the incoming ethanoic acid as nucleophile. In the bromination of alkenes, a broadly similar function of the double bond is observed during formation of a bromonium ion and its subsequent conversion to the *trans*-dibromide (see Section 4.7).

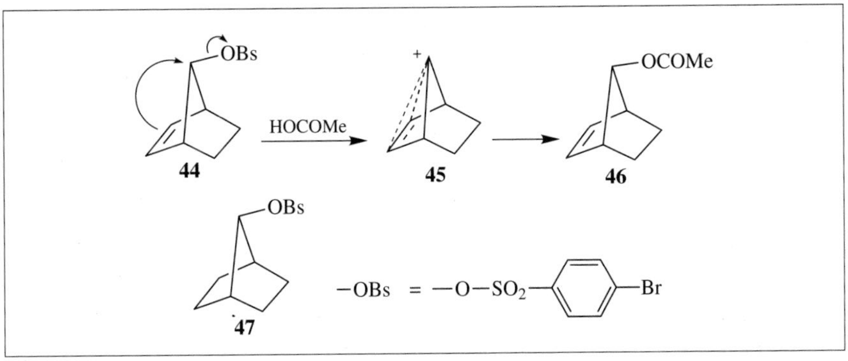

Scheme 7.9

Evidence of a kinetic nature points to involvement of the double bond as shown in Scheme 7.9. The rate constant for the acetolysis of **44** is *ca.* 10^{11} greater than that for **47**, because of the involvement of the electrons of the π-bond in facilitating the ionization of **44**. A small note of caution is in order here: since the C–C double bond is shorter than its single bond counterpart, the free energies of **44** and **47** will be different. However, any difference is relatively small, and will not account for the observed rate constant ratio.

The much greater than expected rate constant for acetolysis of **44** (note

A suitably oriented π bond can massively increase the rate at which the leaving group departs, and participate in formation of a delocalized carbocation that enables the acetolysis to occur with complete overall retention.

that the rate-limiting step is ionization of **44**) together with the net retention of configuration are brought about by the optimally positioned π-bond. For this reason, rate constant enhancement by **44** is probably the most dramatic example attributed to a π-bond.

Worked Problems

Q1. Enantiomerically pure **48**, prepared from **49**, with $[\alpha]_D^{20} =$ –106.1 (neat), underwent reaction with MeS⁻ in ethanol to give the starting compound **49**, which now had $[\alpha]_D^{20} =$ –17.2 (neat). The reaction showed the clean second-order kinetics expected of a bimolecular reaction. Explain (1) how **48** might be made from **49**; (2) the partial racemization of **49** after reaction of the salt **48** with MeS⁻.

A. The salt **48** can be made from **49** by methylation with MeOTs in an S_N2 reaction. When partial racemization is involved, an S_N1 reaction is suspected. However, this is ruled out by the observation of second-order kinetics with a very strong nucleophile. Lists of leaving groups in S_N2 reactions usually include $^+SMe_2$; however,

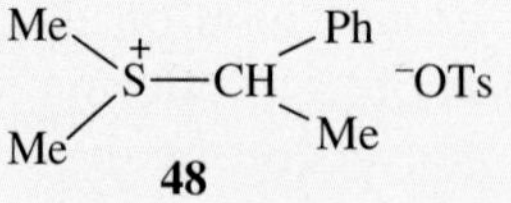

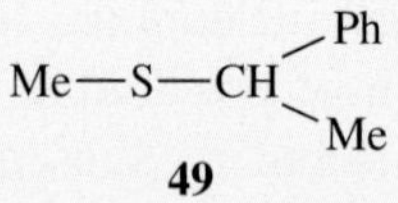

any two alkyl groups as in $^+SR^1R^2$ will suffice. Dimethyl sulfide, Me_2S, can be expelled from **48** to give **49**, as shown in Scheme 7.10. This reaction proceeds with inversion of configuration at the stereogenic carbon. Alternatively, the nucleophile can attack either of the methyl groups bonded to sulfur (Scheme 7.11), and this reaction also gives **49**, but now with retention of configuration. The leaving group is now MeSCH(Me)Ph. The polarimetric result can be explained by: (i) 42% attack by MeS⁻, as in Scheme 7.10; (ii) 58% attack by MeS⁻, as in Scheme 7.11.

Verify these ratios for yourself from the polarimetric data.

The above ratios could be confirmed by use of MeS⁻ labelled with the isotope ^{34}S, and subsequent mass spectroscopic analysis of the products. Note that counterion ⁻OTs in **48** is much less nucleophilic than MeS⁻, and does not interfere.

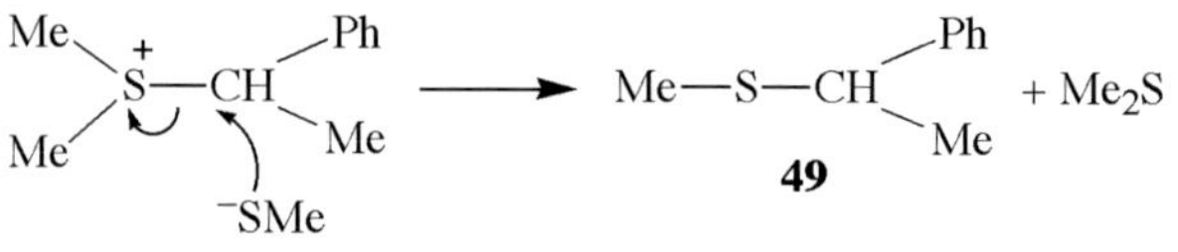

Scheme 7.10

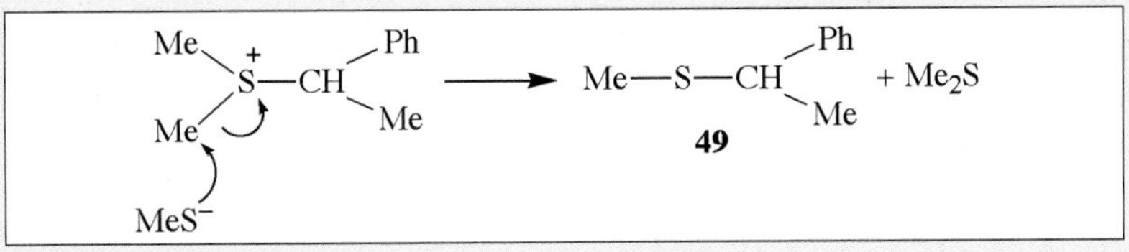

Scheme 7.11

Q2. Explain why ester **50** reacts with lithium iodide in pyridine to give compound **51**, in which only one ester group has been cleaved.

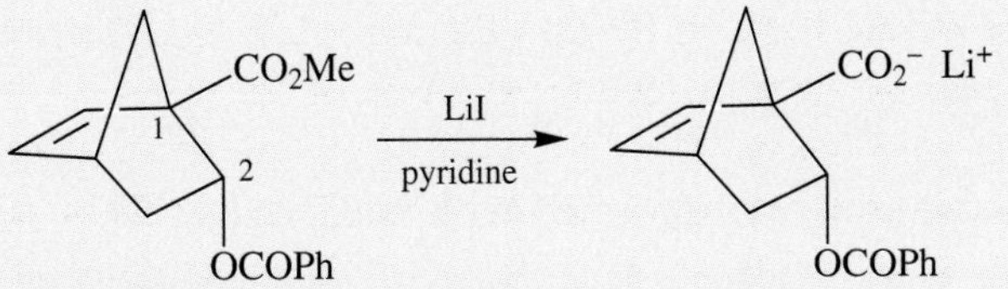

A. The bridgehead methoxycarbonyl group is readily cleaved in a reaction, halolytic fission, that is also an S_N2 reaction at the methyl group. The other ester is more difficult to cleave, because access to the *exo* face of **50** is significantly more difficult on steric grounds. Inspection of space-filling models shows this more clearly.

Summary of Key Points

- Concerted nucleophilic substitution at saturated carbon, the S_N2 reaction, is always accompanied by inversion of configuration. The intramolecular version occurs readily if the reaction is exocyclic (Figure 7.2a), but not if it is endocyclic (Figure 7.2b).
- The S_N2' reaction proceeds with variable stereochemistry according to the particular case; *anti* stereochemistry is associated with RS^- nucleophiles.
- A planar carbocation is formed in the S_N1 reaction of acyclic substrates. Excess inversion rather than the expected racemic product is found when the leaving group shields one face of the carbocation.
- The S_N1 reaction occurs with retention when the leaving group is located at a bridgehead; the more 'tied-back' the structure, the slower the rate constant for ionization.

Problems

7.1. Explain why conversion of **52** to the salt **53** proceeds in good yield with LiSPr.

$$Pr_3C-\underset{\underset{O}{\|}}{C}-OMe \qquad Pr_3C-CO_2^-\ Li^+$$

$$\textbf{52} \qquad\qquad \textbf{53}$$

7.2. Explain why in alkaline solution **54** can be converted into **55**, whereas its diastereoisomer **56** cannot.

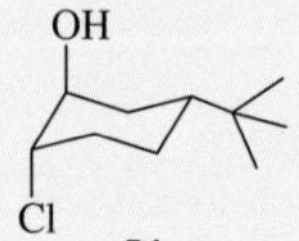

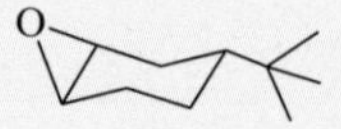

7.3. Show how reaction of **57** with NaCN gives **58**.

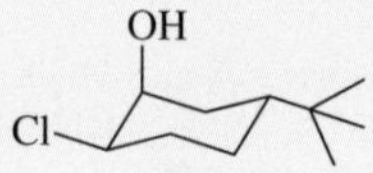

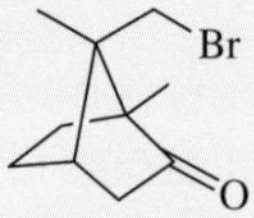

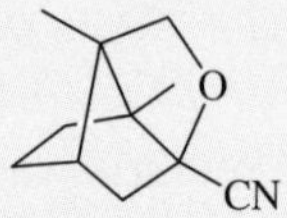

References

1. J. Kenyon, H. Phillips and F. M. H. Taylor, *J. Chem. Soc.*, 1933, 173.
2. E. D. Hughes, F. Juliusberger, S. Masterman, B. Topley and J. Weiss, *J. Chem. Soc.*, 1935, 1525.
3. L. Tenud, S. Farooq, J. Seibl and A. Eschenmoser, *Helv. Chim. Acta*, 1970, **53**, 2059.
4. J. F. King and M. J. McGarrity, *J. Chem. Soc., Chem. Commun.*, 1982, 175.
5. G. W. Kirby, K. W. Bentley, P. Horsewood and S. Singh, *J. Chem. Soc., Perkin Trans. 1*, 1979, 3064.
6. J. Steigman and L. P. Hammett, *J. Am. Chem. Soc.*, 1937, **59**, 2536 (in this paper the term *solvolysis* is used for the first time).
7. S. Winstein and B. K. Morse, *J. Am. Chem. Soc.*, 1952, **74**, 1133.
8. S. A. Sherrod, R. G. Bergman, G. J. Gleicher and D. G. Morris, *J. Am. Chem. Soc.*, 1972, **94**, 4615.
9. P. E. Eaton, *Angew. Chem. Int. Ed. Engl.*, 1992, **31**, 1421.

Further Reading

G. Stork and A. F. Kreft, *J. Am. Chem. Soc.*, 1977, **99**, 3850, 3851.
A. Streitwieser, *Chem. Rev.*, 1956, **56**, 571 (gives a summary of much early work on solvolysis).

8

Prochirality, Enantiotopic and Diastereotopic Groups and Faces: Use of NMR Spectroscopy in Stereochemistry

Aims

By the end of this chapter you should be aware that:

- The CH_2 hydrogens of a prochiral molecule, such as ethanol, are enantiotopic, and are given symbols H_R and H_S
- A molecule such as methyl phenyl ketone, in which the carbonyl carbon is bonded to three different groups, is also prochiral, and the faces of the carbonyl carbon can be described by the symbols *Re* and *Si*
- Reaction of phenylethanone with, for example, lithium aluminium hydride, gives a racemic mixture of enantiomeric alcohols
- The upper and lower faces of (*E*)-butenedioic acid have configurations *Si-Si* and *Re-Re*, respectively
- The CH_2 hydrogens of certain compounds are diastereotopic and they neither show equivalent reactivity nor do they have identical ^1H NMR chemical shifts
- Enantiotopic hydrogens show different reactivity in a chiral environment, and an enzyme is able to distinguish absolutely between them
- Prochiral carbonyl groups and alkenes can show different facial reactivity, exemplified, respectively, by enzymic reduction of

ethanal to ethanol and hydroboration with a dialkylborane to give an enantiomerically enriched alcohol
- When two groups in a molecule that contain respectively H^a and H^b are separated by less than *ca.* 0.4 nm, it is often possible by irradiation of H^a to observe an increase in the integrated signal of the ^1H NMR absorption of H^b. This is the nuclear Overhauser effect. The ease with which the effect is observed for proton H^b decreases along the sequence: $CH^b > C(H^b)_2 > C(H^b)_3$

8.1 Prochiral Molecules, Enantiotopic Groups and Faces

This chapter is concerned in part with prochirality. In order to define terms we consider sp^3 and sp^2 hybridized carbons separately.

8.1.1 Prochirality at sp^3 Hybridized Carbons

Consider the case of an sp^3 hybridized carbon bonded (1) to two groups, typically protons that are chemically identical under achiral conditions, and (2) to two further groups that are identical neither to the groups above, *i.e.* the protons, nor to each other. An example is provided by ethanol (**1**). The two hydrogen atoms in **1** that are bonded to C(1) are termed enantiotopic. This is because replacement of each of these hydrogens, separately, by a group other than the CH_3 and OH groups, already present, will give enantiomers. For the purpose of definition, it is usual to replace the hydrogens bonded to C(1), separately, by deuterium, D. The two molecules so obtained are **2** and **3**. Since compound **2** has (*S*) configuration, the H atom that was replaced by D in **2** is given the symbol H_S in **4**. In the same way, **3** has (*R*) configuration, and in **4**, therefore, the symbol H_R is given to the hydrogen that was replaced by D to give **3**.

In a molecule such as ethanol (**4**), replacement of H_R by D gives a molecule with *R* configuration; this hydrogen is called *pro-R* and the molecule is prochiral. The other methylene hydrogen is *pro-S*. Since separate replacement of H_R and H_S by D gives a pair of enantiomers, H_R and H_S are called **enantiotopic**.

$$C(2)H_3 \text{---} C(1)H_2 \text{---} OH$$

1

2 (*S*)   **3** (*R*)   **4**

A molecule such as **1** is said to be prochiral, and the H_R and H_S atoms in **4** are called *pro-R* and *pro-S*, respectively. The same analysis can be pursued for any molecule of type **5** (with A not the same as B) in which the X groups are enantiotopic. Other examples of prochirality include the methyl groups of dimethyl sulfoxide (**6**) and the oxygens of methyl phenyl sulfone (**7**). The enantiotopic hydrogens in **4** are to be contrasted with those of, for example, dichloromethane (**8**). In **8**, the identity of the two chlorine atoms means that the hydrogens in this molecule are identical under all conditions, and they are said to be **homotopic**. Hanson[1] introduced the term prochiral and related definitions applied to both sp³- and sp²-hybridized carbons.

In a molecule such as CH_2Cl_2, replacement of either hydrogen by deuterium gives the same molecule. Hydrogens in molecules such as CH_2Cl_2 are **homotopic**.

8.1.2 Prochirality at sp² Hybridized Carbons

Organic molecules that contain sp² hybridized carbons can also be prochiral if the substituent pattern is correct. An example is provided by phenylethanone (acetophenone, **9**), which has three different groups attached to the carbonyl carbon. After reduction with, say, LiAlH₄, the product is the chiral alcohol **10** and a 50:50 mixture of enantiomers is observed. The same racemic mixture is obtained from the reaction of the Grignard reagent MeMgBr with benzaldehyde (**11**). In contrast, the product of reduction of propanone (**12**) with LiAlH₄ is propan-2-ol (**13**), which is not chiral.

Reduction of acetophenone (**9**) with LiAlH₄ gives an alcohol with a stereogenic carbon; the alcohol is racemic. The sp²-hybrid carbon is prochiral, and the faces of the carbonyl carbon are enantiotopic.

Therefore, an sp² hybridized carbon bonded to three different groups will become chiral as the result of an addition reaction (which may also be a reduction) that results in formation of a bond between the sp² hybridized carbon and a fourth group that differs from the other three. Such sp² hybridized carbons are called prochiral. This raises an interesting point of nomenclature. A chiral molecule that has four different groups attached to a carbon atom is said to have a **stereogenic centre**, this term being preferred to 'chiral centre'. However, a prochiral molecule is said to have a prochiral centre; the term 'pro-stereogenic', preferred by Helmchen,[2] is not yet widely adopted.

The racemic mixture that results from **9** and LiAlH₄ arises because one enantiomer is formed by attack of LiAlH₄ at the top face of **9**, and the other from attack at the bottom face. It is possible to characterize the faces of the carbonyl carbon of **9** by means of a modification of the Cahn–Ingold–Prelog (CIP) *R/S* (*rectus/sinister*) convention.

Consider the case of phenylethanone, re-drawn in **14**; the CIP sequencing of substituents is O > Ph > Me. One looks down on the carbonyl group from the top face, and since the priorities decrease in an anticlockwise sense, this face is described as *Si* (the first two letters of *sinister*). Looking at the carbonyl carbon from the bottom face, the substituent priorities decrease in a clockwise sense, and so the lower face, as drawn, of the carbonyl carbon is described as *Re* (the first two letters of *rectus*). Previously, the symbols used were *re* and *si*; now, *Re* and *Si* are preferred (see Eliel and Wilen,[3] Helmchen[2] and Aitken and Kilényi[4]). Both Eliel and Wilen and Helmchen suggest that *re* and *si* be used in a different context, which is not relevant here.

The *Re/Si* symbols can similarly be applied to the carbon atoms that make up C=C double bonds. If one considers the alkene **15**, the top face can be described by *Re-Re* and the lower face by *Si-Si*. (*E*)-Butenedioic acid (fumaric acid, **16**) has a top face configuration (as drawn) of *Si-Si*, and the lower face has a configuration of *Re-Re*.

A molecule such as **14** is said to have two enantiotopic faces. Attack by, say, LiAlH$_4$ at the *Re* face will produce one enantiomer, and attack at the *Si* face will produce the other. There is no correlation between the configuration of the carbonyl face and that of the tetrahedral product after addition of a reagent. It all depends on the priority rating of the added reagent with respect to the groups in the original carbonyl compound.

Consider attack at the lower, *Re*, face of the carbonyl carbon of **14**. When the reagent is LiAlH$_4$, the product is **17**, which has (*S*) configuration. In contrast, attack on the same *Re* face of **14** by EtMgBr gives the tertiary alcohol **18**, which has (*R*) configuration.

When a carbonyl carbon is attached to two identical groups as in methanal (formaldehyde) or propanone (acetone), attack at the top face is identical to attack at the bottom. Therefore attack at either face (by an achiral reagent) gives a product that is achiral. The carbonyl-containing group is called homotopic.

Make a model of propanone to confirm the above statements. Use the model to verify that propanone (as an example of a homotopic carbonyl group) contains: (1) a two-fold axis of symmetry that passes through the carbon and oxygen of the carbonyl group; (2) two planes of symmetry that pass through the carbon and oxygen of the carbonyl group. These planes of symmetry are at right angles to each other; one is in the plane of the p-orbitals that form the π bond, and the other is in the nodal plane of this p-orbital.

8.2   Diastereotopic Compounds

If one takes a molecule such as **19**, in which C(2) is a stereogenic centre, and separately replaces H^a with D, and then H^b with D, one obtains **20** and **21**, respectively. The compounds **20** and **21** are diastereoisomers, and so the protons H^a and H^b in **19** are termed **diastereotopic**. This name gives an indication that the reactivity of H^a and H^b toward achiral reagents is different.

In compound **19**, hydrogens H^a and H^b are adjacent to a stereogenic centre, and their separate replacement by D gives diastereoisomers. These hydrogens are **diastereotopic**.

$$\begin{array}{ccc} \text{19} & \text{20} & \text{21} \end{array}$$

In addition, as protons H^a and H^b in **19** are chemically inequivalent, they necessarily, in principle, experience different magnetic shielding, and so have different chemical shifts in achiral solvents. Since H^a and H^b are magnetically non-equivalent, one can observe a coupling constant $^2J_{H^aH^b}$ between these nuclei. However, the shielding differences between H^a and H^b are generally small, and there are instances in which it is zero. When this happens, H^a and H^b have the same chemical shift and, of course, no coupling is now observed between these protons. Where diastereotopic protons show the same chemical shift, they are said to be accidentally equivalent or **isochronous**, and where they have different chemical shifts the protons are described as **anisochronous**.

Diastereotopic protons are **anisochronous**, and have different chemical shifts.

In the similar way that **19** has diastereotopic protons, it is possible for molecules to have diastereotopic faces. Examples of these are provided by the acyclic ketone **22**, as well as by camphor (**23**). These two molecules both have one or more stereogenic centres.

The effect of chirality on the NMR spectrum is also discussed in Gunther.[5]

A carbonyl carbon located in the same molecule as a stereogenic centre is diastereotopic. Reagents will approach *one face* of such a carbonyl carbon more easily than they will approach the other.

8.3 Reactivity Difference of Enantiotopic Hydrogens and Faces in Enzyme-mediated Reactions

Enantiotopic hydrogens react at identical rates in an achiral environment. In order to bring about a difference in their reactivity, an external chiral influence has to be involved. This can be provided by enzymes, as shown in the following two reactions.

Conventional laboratory oxidation of the monodeuteriated ethanol **24** gives ethanal (acetaldehyde), which contains 50% of the deuterium present in **24** (Scheme 8.1). However, enzymes are chiral, and enzymic oxidation (by yeast alcohol dehydrogenase and NAD$^+$) removes H$_R$ exclusively in the oxidation of ethanol to ethanal; H$_R$ is labelled as 'D' in Scheme 8.2. Clearly, the molecule of ethanol is presented to the chiral enzyme so as to form a unique diastereoisomeric complex in which only the H$_R$ proton can be removed in the oxidative elimination.

Whereas 'normal' laboratory oxidation of ethanol involves loss of either proton from the CH$_2$ group, H$_R$ is lost *exclusively* in the corresponding enzymic oxidation.

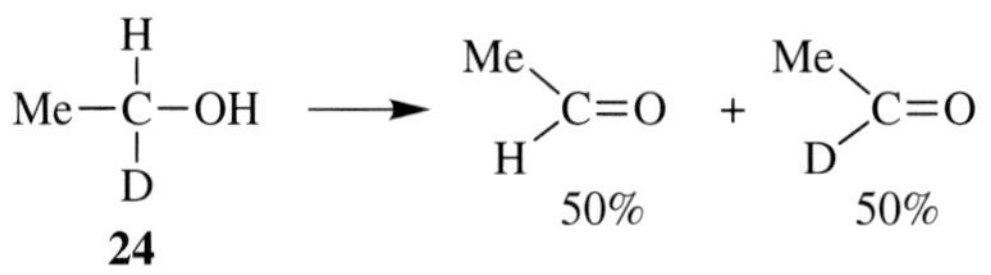

Scheme 8.1

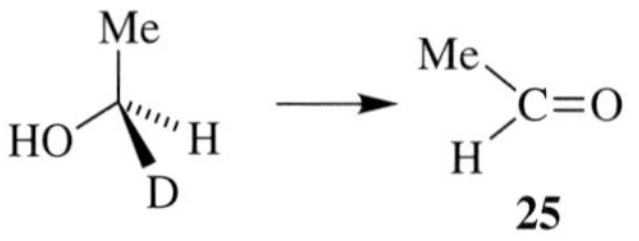

Scheme 8.2

Addition to the two faces of a carbonyl group shows a similar dual behaviour. Firstly, reduction of ethanal (**25**) with LiAlD$_4$ gives ethanol in which the H$_R$ and H$_S$ sites are both populated with deuterium to the extent of 50%. This is because reduction occurs with equal ease at the *Re* and *Si* faces. However, it has been shown that the deuteriated ethanal **26** undergoes reduction with yeast alcohol dehydrogenase and its coenzyme NADH to give exclusively the (*S*) enantiomer **27** (despite their name, dehydrogenases are effective in both oxidations and reductions). Formation of **27** arises because of delivery of hydrogen exclusively to the *Re* face of the carbonyl carbon in the enzyme complex.

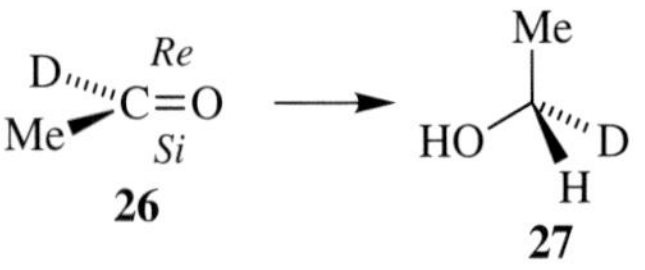

Similarly, (*S*)-phenylmethanol (benzyl alcohol, **29**) is produced exclusively by reduction of deuteriated benzaldehyde **28** by the liver alcohol dehydrogenase–NAD$^+$ reduction system.

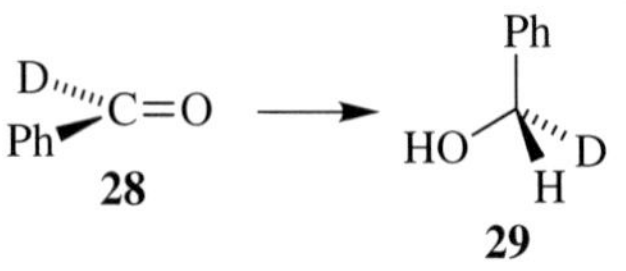

Enzymic reduction of deuteriated ethanal with NADH gives the (*S*)-enantiomer of ethanol because of delivery of hydrogen solely to the *Re* face of the carbonyl carbon of ethanal.

To which face of the carbonyl group of **28** is the new C–H bond formed in giving **29**?

With this background, one can ask: what are NAD$^+$ and NADH? These are complex chiral molecules, and the part structure of NAD$^+$ is shown in **30**. NAD$^+$ contains two ribose sugar units, one of which is shown in order to demonstrate that the aromatic ring is prochiral. A more condensed structure is shown in **31** and the product of reduction of NAD$^+$ is **32**, known as NADH. In **32**, the two hydrogens H^a and H^b are diastereotopic because R is chiral. Biological oxidations of organic compounds that involve NAD$^+$ and an enzyme catalyst always proceed with completely diastereoselective transfer of hydrogen (with its bonding electrons) to one face of **30**.

The hydrogens H^a and H^b in NADH are diastereotopic, and in the presence of an enzymic catalyst, transfer of H^a, or H^b depending on the reaction, takes place exclusively in reductions.

30   **31**   **32**

The face to which the hydrogen is delivered is a function of the reaction. Oxidation of glucose with NAD$^+$ and enzymes from the liver gives **32**, in which the transferred hydrogen is located exclusively at H^b. On the other hand, enzymic oxidation of ethanol, now using NAD$^+$ and enzymes from yeast, gives **32**, in which the transferred hydrogen is located exclusively at H^a.

8.4 Non-enzymatic Selectivity in Addition to C=O and C=C Bonds

In the previous section we have seen that enzyme-mediated reductions of carbonyl groups proceed with complete facial selectivity. Enzymes are complicated chiral molecules and synthetic organic chemistry is challenged to come consistently close to matching their selectivity. What we do here is merely point to some of the general directions that are followed in order to achieve selectivity.

Within the context of ketones and alkenes as substrates one must do either of the following:

(1) Set up a prochiral ketone (or alkene) with structural features that direct the reagent preferentially to one face. In the product, it will probably be necessary to have present in the molecule the features that made one face more attractive to the incoming reagent. Clearly, for purposes of synthesis this may not be desirable

(2) Alternatively, one can structure the reagent so that it attacks the double bond of either a carbonyl or an alkene at one face preferentially. Examples of both reaction types, one with a carbonyl and one with an alkene, are now given.

In the case of camphor, shown as the (1*R*) enantiomer in **23**, the *exo* face is *Re*, and is shielded to a greater extent than the lower *endo* face, which is *Si*. Note that in the enantiomer of **23** the upper face is still *exo*, but is now described as *Si*, and the lower *endo* face is *Re*. In **23**, the greater shielding of the *exo* face suggests that reduction with LiAlH$_4$ involves preferential delivery of hydride to the *endo*, *Re*, face. It is found that the major product **33** has the OH group *exo* (2*R*) and the ratios of **33:34** are 75:25. A more bulky reducing agent, LiAl(OBu$^t_3$)H, is more selective still, and gives a 99:1 ratio of **33:34**.

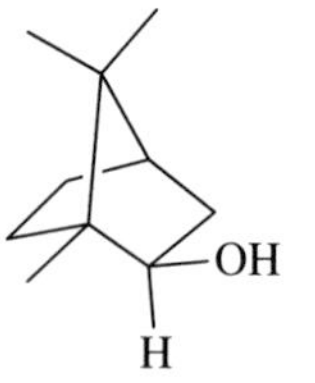

If a prochiral ketone or alkene does not have any significant inbuilt structural characteristics that induce a reagent to attack one face or the other, it is still possible to achieve a high degree of facial selectivity, and so one can obtain a product of high enantiomeric purity. An example is provided by the asymmetric hydroboration of prochiral alkenes.

The background to this method is provided by the hydroboration of alkenes with diborane, B$_2$H$_6$ (Section 4.8.2). A borane from the very bulky chiral alkene α-pinene (**35**) is **36**, which is of general form R$_2$BH. This borane is monomeric, and is formed by *cis* addition of diborane to the double bond in each of two α-pinene molecules. In this chiral diborane, the bulk of α-pinene prevents formation of a trialkylborane of type R$_3$B. However, addition of R$_2$BH **36** (Scheme 8.3) to a *smaller* prochiral alkene occurs preferentially to one face, and is the basis of a synthesis of alcohols with a high degree of enantiomeric purity.

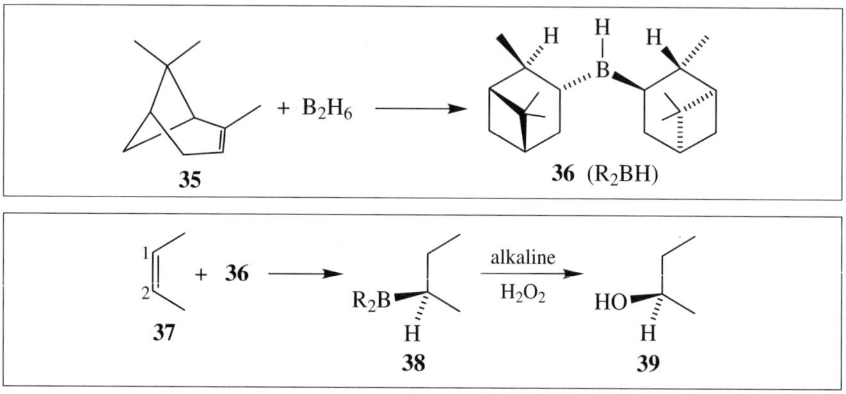

33   **34**

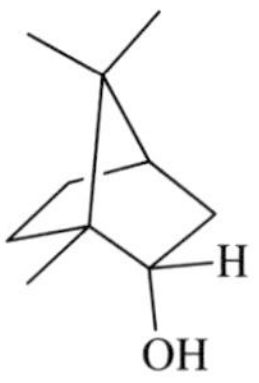

One successful laboratory procedure to form enriched alcohols from prochiral alkenes is shown with **37** (Scheme 8.3). The key reaction is a hydroboration reaction with R$_2$BH, in which R is derived from an enantiomerically pure natural product, α-pinene.

Scheme 8.3

For example, (*Z*)-but-2-ene (**37**) reacts with the chiral dialkylborane **36** to produce **38** as the result of the addition of the borane to the upper (*Re*) face of C(2), or the equivalent lower (*Re*) face of C(1). Conventional work-up of the borane **38** with alkaline hydrogen peroxide (Scheme 8.3) gave **39**. This was predominantly of R configuration; an enantiomeric excess of 87% (or an enantiomeric ratio of 14.4:1) was observed. This is a remarkably high figure when one considers the relatively modest bias provided by the hydrogen *versus* methyl differential on (Z)-but-2-ene.

Another example is provided by reaction of phenylethene (styrene, **40**) with the reagent **41** in methanol. Reagent **41** contains a chiral aromatic group (Ar*) and selenium is also bonded to the trifluoromethylsulfonate (triflate) group, OSO_2CF_3, which is similar to, but better than, tosylate as a leaving group. Reaction takes place as indicated in Scheme 8.4, to give the chiral seleniranium ion intermediate **42** (see Wirth *et al.*[6]). This ion is of the same general form as a bromonium ion (Section 4.7). Ion **42** then reacts with methanol, as indicated, and the product **43** is formed with a diastereoisomeric ratio (dr) of 96:4 in favour of the isomer **43**.

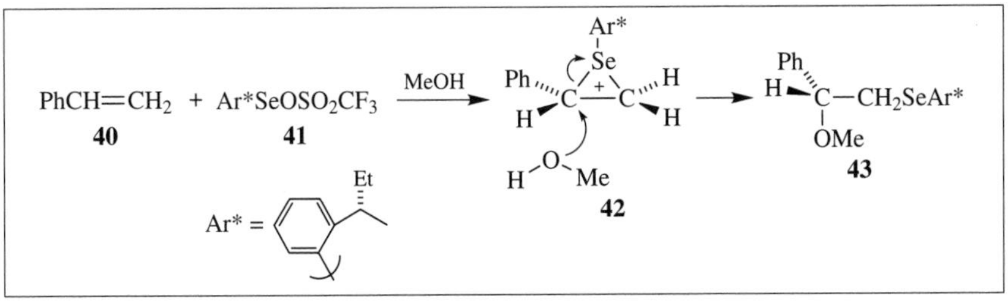

Scheme 8.4

8.5 Stereochemical Information from NMR Spectroscopy

Nuclear magnetic resonance (NMR) spectroscopy can provide stereo-chemical information, both conformational and configurational, as described in the following sections. A working knowledge of NMR spectroscopy is assumed.

8.5.1 Vicinal Coupling Constants

The E/Z ratio in a mixture of disubstituted alkenes can readily be determined from the relevant integrated signals once the absorptions have been assigned. We can use the knowledge that when substituents are more electronegative than hydrogen, vicinal coupling constants in *cis*- or (Z)-alkenes of type **44** are in the range $^3J_{HH} = 7–11$ Hz, whereas the corresponding values for *trans*- or (E)-alkenes such as **45** are given by $^3J_{HH} = 12–18$ Hz (see Friebolin[7]). The magnitudes of the coupling constants vary slightly with electronegativity of the substituents, but the above ^3J values for (E) and (Z) disubstituted alkenes do not overlap, and so assignments can be made with confidence.

$$
\begin{array}{cccc}
\underset{\text{Cl}}{\overset{\text{H}}{\diagdown}}\text{C}=\text{C}\underset{\text{CO}_2\text{H}}{\overset{\text{H}}{\diagup}} &
\underset{\text{Cl}}{\overset{\text{H}}{\diagdown}}\text{C}=\text{C}\underset{\text{H}}{\overset{\text{CO}_2\text{H}}{\diagup}} &
\underset{\text{F}}{\overset{\text{H}}{\diagdown}}\text{C}=\text{C}\underset{\text{H}}{\overset{\text{F}}{\diagup}} &
\underset{\text{F}}{\overset{\text{H}}{\diagdown}}\text{C}=\text{C}\underset{\text{F}}{\overset{\text{H}}{\diagup}} \\
\mathbf{44} & \mathbf{45} & \mathbf{46} & \mathbf{47}
\end{array}
$$

The same order holds when the vicinal coupling constants are between a proton and fluorine. This is shown in the *cis* coupling constant $^3J_{HF}$ between –4 and 20 Hz in molecules such as **46**, and the corresponding *trans* coupling constants $^3J_{HF}$ between 20 and 100 Hz in, for example, **47** and related molecules.

In saturated cyclic molecules that are rigid, information about structure can be obtained from vicinal coupling constants $^3J_{HH}$. In cyclohexane, for example, there are three vicinal coupling constants and the values were determined at –103 °C, at which temperature ring inversion is very slow. To simplify matters, the compound used was labelled with deuterium at four adjacent carbons and is named 1,1,2,2,3,3,4,4-octadeuteriocyclohexane (see Garbisch and Griffith[8]).

> Draw this compound in a chair conformation.

Vicinal coupling constants in alkenes, and in conformationally fixed cyclohexanes, provide information on stereochemistry. These coupling constants in saturated molecules vary as a function of the cosine of the dihedral angle concerned.

The proton–proton coupling constants were measured and found to be: $^3J_{Hax–Hax} = 13.1$ Hz; $^3J_{Hax–Heq} = 3.65$ Hz; and $^3J_{Heq–Heq} = 2.96$ Hz. More generally observed values for conformationally rigid cyclohexanes are in the ranges $^3J_{Hax–Hax} = ca.$ 10–13 Hz and $^3J_{Hax–Heq}$ and $^3J_{Heq–Heq} = ca.$ 2–5 Hz.

Karplus[9] recognized that vicinal coupling constants between protons H^a and H^b are a function of the dihedral angle H^a–C–C–H^b, denoted by ϕ, and this angular dependence is shown in Figure 8.1. The value of $^3J_{HH}$ is slightly higher at 180°, and the graph in Figure 8.1 is closely reproduced by the Karplus (sometimes referred to as the Karplus–Conroy) equations (1) and (2) (see Breitmaier[10] and Haasnoot *et al.*[11]). In an unknown molecule that has a rigid ring structure of, say, six carbons, it is possible to determine dihedral angles to within approximately 5° from a knowledge of vicinal coupling constants, if electronegativity effects are taken into account.

$$
^3J_{HH} = A\cos^2\phi - 0.28 \tag{1}
$$

where $A = 8.5$ for $\phi < 90°$.

$$
^3J_{HH} = B\cos^2\phi - 0.28 \tag{2}
$$

where $B = 9.5$ for $90° < \phi < 180°$.

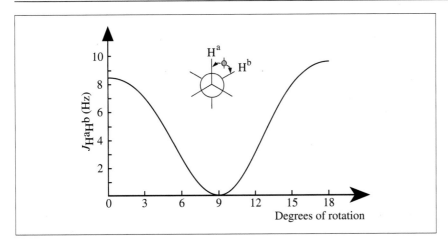

Figure 8.1 Karplus curve showing dependence of the vicinal coupling constant $^3J_{H^a-C-C-H^b}$ (in Hz) on the dihedral angle ϕ

8.5.2 Diastereomeric Compounds and Interactions: Enantiomer Recognition

Analogously, one can determine the relative amounts of two diastereoisomers, of general formula **48**, by integration of a pair of comparable signals in a mixture. The data can be expressed as either (1) a diastereomeric ratio or (2) a diastereomeric excess in a similar manner to enantiomeric excess. The diastereomeric excess (de) of one diastereoisomer, X, over another, Y, is given in equation (3). Although one can determine the de values from NMR spectroscopy, no information on the particular configurations is obtained.

$$de = \frac{\%X - \%Y}{\%X + \%Y} \qquad (3)$$

In a normal NMR experiment in, say, $CDCl_3$ or CCl_4, it is not possible to distinguish enantiomers or estimate enantiomeric purities. However, by the use of three separate techniques that all involve some diastereomeric component, it is possible to determine enantiomeric purities of, for example, alcohols, esters, ketones and amines.

Firstly, one can use a 'chiral solvating agent' (CSA) in which one relies on diastereomeric interactions, or complexations, to give a chemical shift difference for the two enantiomers. The method was first used to demonstrate ^{19}F anisochrony (Section 8.2) for the enantiomers of 2,2,2-trifluoro-1-phenylethanol (**49**) in the presence of (S)-(−)-1-phenylethylamine (**50**) (see Pirkle[12]). CSA-induced chemical shift differences in ^1H NMR spectra are generally small.

The second technique is to use a **chiral lanthanide shift reagent** (CLSR). Addition of certain lanthanide chelates, most commonly those of europium(III), to a solution of an organic compound in $CDCl_3$ or

Enantiomers can give separate signals in their NMR spectra: (1) with a chiral solvating agent (CSA); (2) with a chiral lanthanide shift reagent (CLSR); (3) in the case of alcohols, reaction with one enantiomer of a carboxylic acid, *e.g.* Mosher's acid. With other cited acids it is sometimes possible to assign alcohol configuration directly from the ^{1}H NMR spectrum.

51

CCl_4 brings about large downfield shifts of proton absorptions (see Hinckley[13]). The shifts are known to be dependent on the concentration of europium chelate, and addition of excess chelate, or use of a high field such as that at 500 MHz, gives line-broadened spectra of low precision.

The use of an enantiomerically pure chiral ligand bonded to, say, europium(III), gives the CLSR. Of these, **51**, in which the ligand is a fluorinated camphor derivative, has been extensively used, though analogous complexes of praseodymium and ytterbium have been advocated as preferred alternatives (see Parker[14] and Rinaldi[15]). The purpose of the CLSR is to interact with the two enantiomers in a distinct manner, possibly by formation of weak diastereomeric complexes. This differential interaction together with the shift properties of the CLSR enables the ^{1}H absorptions of the respective enantiomers to be integrated with certainty.

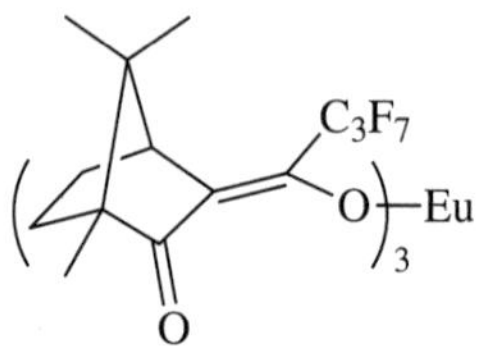

Thirdly, in the case of completely or partially racemic alcohols, it is possible to create diastereomeric esters by reaction with one enantiomer of a chiral carboxylic acid. One acid that gives consistently good results is Mosher's acid, 3,3,3-trifluoro-2-methoxy-2-phenylpropanoic acid (**52**) (see Dale *et al.*[16]). With a chiral alcohol, *e.g.* 1-phenylethanol, a pair of diastereoisomers, **53** and **54**, is formed (Scheme 8.5).

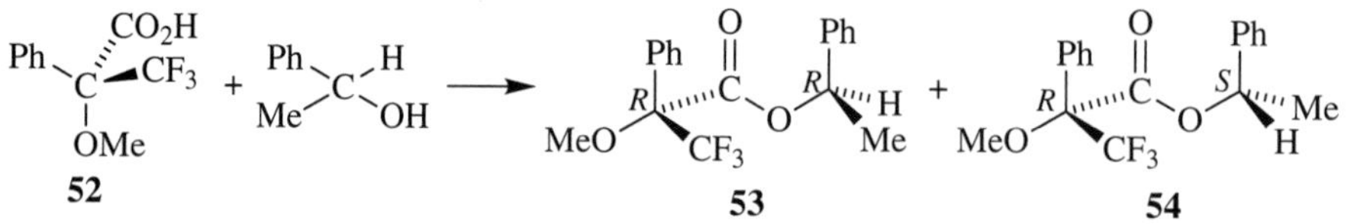

Scheme 8.5

55

An advantage of this ester is that the diastereoisomer ratio can be measured in two ways: (1) from the ^{1}H NMR spectrum, *e.g.* by integration of the methoxy proton absorptions of **53** and **54**; (2) by integration of the ^{19}F absorptions of the CF_3 groups of this pair of compounds. With the aid of esters **53** and **54** it is possible to determine the ratio of two enantiomeric alcohols in a mixture. However, assignment of absolute configuration to the alcohols can be problematic.

Other carboxylic acids such as **55** have been used to make corresponding diastereoisomeric esters. The choice of aromatic substituent in **55** is made on the basis of the greater anisotropy of the three fused aromatic rings in **55** with respect to phenyl. It has recently been possible to assign configuration to a pair of enantiomeric alcohols directly from the ^{1}H NMR spectra of their esters with **55**. The discussion goes beyond the scope of this text, but for details the work of Fukushi *et al.*,[17] Takahashi *et al.*[18] and Seco *et al.*[19] should be consulted.

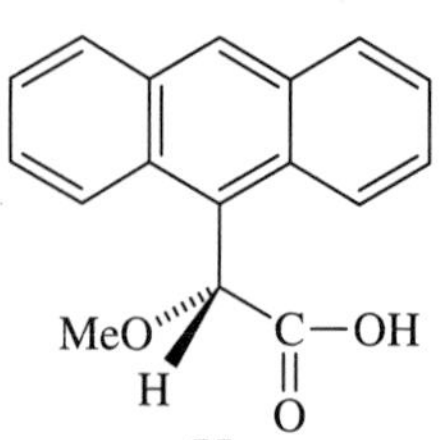

8.5.3 Nuclear Overhauser Effect

A commonly observed interaction of two magnetically non-equivalent nuclei is the coupling constant J. The coupling mechanism is brought about by electronic interactions that pass through intervening bonds. A through-space interaction is considered to account for long-range couplings (see Figure 8.2 on page 153, in which H^{3exo} shows a doublet as a result of coupling with H^{5exo}). It is possible to have a further through-space interaction between two non-equivalent nuclei, H^a and H^b, in the same molecule and for these nuclei to be well separated in terms of numbers of intervening bonds. This interaction can be shown by irradiation of the sample at the resonance frequency of, say, H^a, and observation of the integrated intensity of H^b. One can observe an enhanced integral for H^b when the molecule in question has a relative molecular mass (M_r) of less than approximately 300. It is also helpful if the nuclei H^a and H^b are closer than 0.4 nm, although slightly greater distances are possible in the case of more slowly tumbling biopolymers. The enhanced integration outlined above is known as the **nuclear Overhauser effect** (NOE). The effect has its origins in the dipolar relaxation of the proton H^b.

It is possible to observe NOEs for protons that are bonded to sp^3, sp^2 or sp carbons. The effect is best observed for methine ($R_3C–H$) protons, though it can be demonstrated for methylene (R_2CH_2) protons also. It is less easy to observe for methyl (CH_3) protons, because methyl protons contribute mutually to the dipole–dipole relaxation of each other. A methine proton, being bonded to three carbons that do not assist in dipole–dipole relaxation, shows an NOE more readily. Therefore, in the case of a closely separated proton, H^a, and a methyl group $C(H^b)_3$ in a molecule, irradiation at the methyl proton frequency results in an enhancement of the H^a absorption, whereas irradiation of H^a will not usually result in a meaningful enhancement of the methyl group absorption. However, as indicated later in this section, it is possible to obtain small NOEs for certain methyl groups.

In the case of ^1H NMR spectroscopy, the maximum signal enhancement is 50%, though typical values are in the range 8–15%. The magnitude of NOEs falls off rapidly with separation, r, of the nuclei H^a and H^b, and an r^{-6} dependence has been reported (see Bell and Saunders[20,21]), and is also predicted theoretically.

One of the largest NOE effects was shown by **56**; here the methine proton H^a experiences a 45% NOE on irradiation of H^b, and *vice versa*. In the case of **57**, the NOE of the proton attached to C(8) is 8% when the protons of the C(20) methyl group are irradiated. Models indicate that the relevant hydrogens here are separated by *ca.* 0.3 nm, on average (see Anet and Bourn[22]).

In the case of two non-equivalent nuclei in the same molecule, it is possible to observe an enhanced integral for H^b if the resonance frequency of H^a is irradiated. The effect works best when H^b is a methine proton ($R_3C–H$), and works least well for methyl protons. The separation of H^a and H^b should be less than 0.4 nm This nuclear Overhauser effect (NOE) has a maximum of 50% for H/H interactions; compound **56** shows 45%. NOEs are very useful for structure determination.

56   **57**

Make a model of **57** to show the proximity of the C(20) methyl group and the C(8) methine hydrogen.

58

The conformation of 2,4-dioxabicyclo[3.3.1]nonane (**58**) was investigated in [^{2}H$_6$]benzene. NOE enhancements of about 5% were found for both H$^{3\beta}$ and H^{9syn} when the other was irradiated, and absence of a corresponding effect for H$^{3\alpha}$ and H$^{7\alpha}$ indicates that the predominant conformation is shown in **58**; this is called a chair-boat conformation because one six-membered ring is in the chair form, and the other is in the boat form (see Peters *et al.*[23]).

It has been claimed that the parent hydrocarbon corresponding to **58**, bicyclo[3.3.1]nonane, exists in a twin-chair conformation. The basis for this proposal came from the ^{1}H NMR spectrum of the hydrocarbon (see Peters *et al.*[24]) and also from the infrared (IR) spectrum, which showed a distinct band at *ca.* 1490 cm^{-1}, indicative of a strong transannular interaction of the methylene groups at C(3) and C(7) (see Eglinton *et al.*[25]). Although results from IR spectra are probably less incisive than those from NOE experiments, they are mutually supportive.

NOEs can be revealed by computer subtraction of the 'normal' spectrum from that recorded while a particular proton is being irradiated. The resulting difference spectrum shows only the enhanced signals (as well as that of the irradiating frequency). Even though difference spectra involve a decrease in sensitivity, they permit detection of NOE enhancements of the order of 1%. In particular, in the case of closely separated intramolecular methine and methyl groups, it is now possible to observe a NOE enhancement of the methyl proton signal, as well as that of the more readily observed methine proton. At present, a 1D-NOE experiment using a pulsed field gradient selective excitation (PFGSE) avoids problems of sensitivity loss but still gives a 'difference' presentation (see Stonehouse *et al.*[26] and Stott *et al.*[27]). Here the NOE signal is generated as usual by distance-dependent H/H cross relaxation. The PFGSE spectra of 3-*endo*-bromo-4-chlorocamphor (**59**, Figure 8.2) reveal the usefulness of the technique in stereochemical assignments. For historical reasons the signals are nevertheless sometimes referred to as 'enhancements'. However, inspection of Figure 8.2 spectra a and b (δ 4.64), spectra a and c (δ 1.74 and δ 2.09) and spectra a and d (δ 0.966) suggest that this term might not always be appropriate.

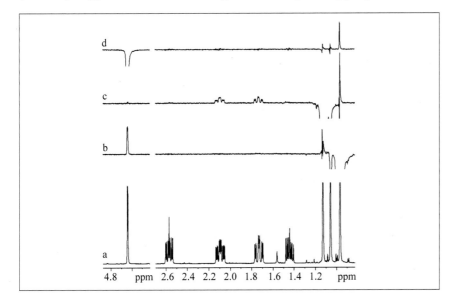

Figure 8.2 PFGSE ^{1}H NMR spectrum of a 0.1 M solution of 3-*endo*-bromo-4-chlorocamphor (**59**) in CDCl$_3$ at 400 MHz. (a) Normal spectrum. (b) After irradiation of the C(8) methyl protons at δ 0.966. This results in enhancement of the H^{3exo} signal at δ 4.641. Signals for the methylene protons are absent. (c) After irradiation of the C(9) methyl protons at δ 1.128. NOE signals for the C(5) and C(6) *exo* methylene protons are observed. Signals for the *endo* methylene protons and the H^{3exo} proton are absent. (d) After irradiation of H^{3exo} at δ 4.461 a NOE signal for the C(8) methyl proton is observed at δ 0.966

It is worth mentioning that larger molecules with M_r greater than approximately 1000 give negative NOEs, and this reduction in signal intensity is found also with increased solvent viscosity, *e.g.* with [^{2}H]$_6$-DMSO.

Texts on the Overhauser effect by Neuhaus and Williamson, Noggle and Schirmer (see Further Reading) and Freeman[28] should be consulted for more details. That by Freeman gives a graphic ski-slope analogy to some of the mechanics of NOEs. Henmann *et al.*[29] give an example of the use of NOEs in determining the stereochemistry of a compound.

Worked Problems

Q1. In the following molecules, **60–65**, indicate whether the hydrogens marked H^a, H^b are homotopic, enantiotopic or diastereotopic.

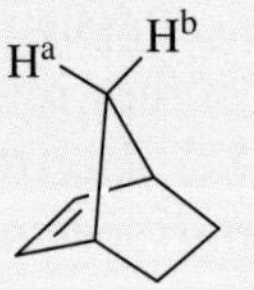

63 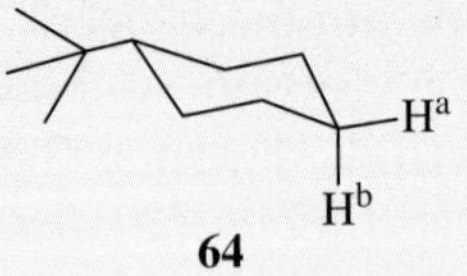   **64** 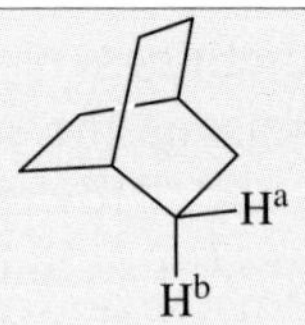   **65**

A.   Compound **60** is of the same substitution pattern as dichloromethane, and the hydrogens H^a and H^b are homotopic. Compound **61** has the same pattern as ethanol; here H^a and H^b are enantiotopic. In compound **62**, H^a and H^b are homotopic; here as in other homotopic examples, H^a and H^b lie astride a symmetry plane. In compound **63**, H^a and H^b are diastereotopic. Separate replacement of H^a and H^b by D in **63** gives a pair of diastereoisomers (that is stereoisomers that are not enantiomers). In compound **64**, H^a and H^b are diastereotopic; this can be confirmed by separate replacement of H^a and H^b by D, to give a pair of diastereoisomers. Note that previously (Sections 6.3 and 6.5.2) we have shown that *cis* and *trans* 1,3-disubstituted cyclobutanes and *cis* and *trans* 1,4-disubstituted cyclohexanes constitute a pair of achiral diastereoisomers. In compound **65**, H^a and H^b are enantiotopic. In fused-ring molecules, substituents are often drawn where H^a/H^b are located in **65**; however, anti-clockwise rotation by 120° around an axis through the bridgehead carbons puts H^a and H^b in a position where their enantiotopicity can be seen more easily.

Q2.   Are the faces of the carbonyl groups in compounds **66–71** homotopic, enantiotopic or diastereotopic?

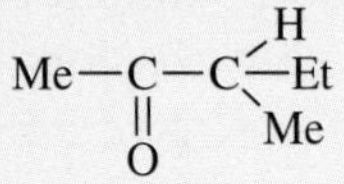

66    **67** 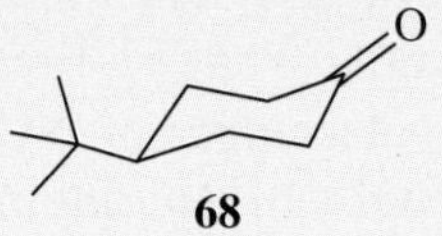   **68**

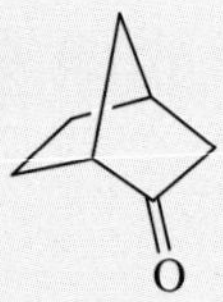

69 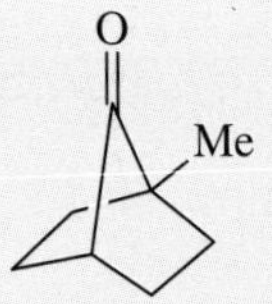   **70** 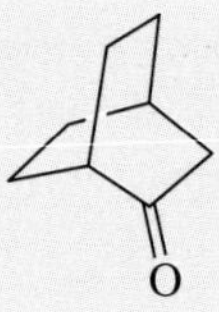   **71**

A.   In compound **66** the carbonyl group is diastereotopic because of the adjacent stereogenic centre at C(2). Therefore addition of, say, $LiAlH_4$ to **66** will give a pair of diastereoisomers. There is one exception: if MeMgBr is added to **66**, the product is **72**, which contains only one stereogenic centre. However, in **72** the two pairs of methyl hydrogens on C(2) are themselves diastereotopic by virtue of the same adjacent stereogenic centre at C(3).

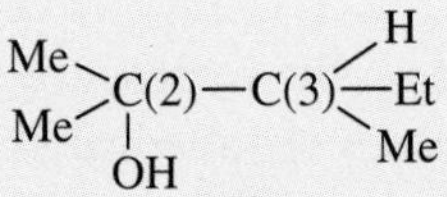

72

Compound **67** has a homotopic carbonyl group; an identical product is formed when a reagent ($LiAlH_4$) attacks the carbonyl group from either face. Compound **68** has a diastereotopic carbonyl group. Attack by $LiAlH_4$ at the two faces of the carbonyl group gives two diastereoisomers (in this case the diastereoisomers are not chiral). In compound **69** there are two stereogenic centres (the bridgehead carbons) and since the unsubstituted bridges are of unequal length, the carbonyl group is diastereotopic. Ketone **70** has an enantiotopic carbonyl group, as does **71**. From both these ketones one obtains, for instance, two enantiomeric alcohols after reduction with $LiAlH_4$.

Q3. With reasons, state whether you would expect the protons H^a, H^b in **73** to be isochronous.

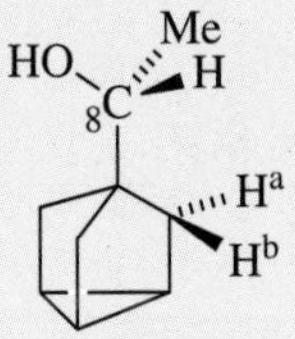

73

A. In **73** there are three equivalent sets of H^a, H^b protons on the three methylene groups (only one set is shown). The stereogenic centre at C(8) makes H^a and H^b non-equivalent. Note that in this, and other related examples, it is perfectly acceptable to work with a racemic sample. If the configuration at the stereogenic centre is inverted from that shown, then the chemical shifts of H^a and H^b are reversed. The appearance of the spectrum is unchanged (see Morris et al.[30]).

Q4. Would you expect the protons H^a and H^b in **74** to be enantiotopic?

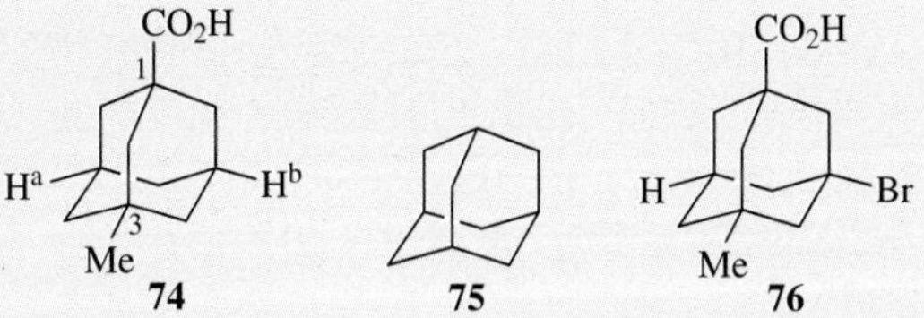

A. Together with methane, the hydrocarbon adamantane (**75**), the parent structure of **74**, has a tetrahedral skeleton. Therefore derivatives of adamantane that have four different substituents at the bridgehead positions should be chiral; this has been demonstrated for **76** (see Hamill and McKervey[31]). It follows that in **74**, H^a and H^b are enantiotopic, and indeed **74** has a plane of symmetry through C(1) and C(3). A parallel situation exists in ethanol, which has a plane of symmetry through two carbon atoms and oxygen. Although H^a and H^b in **74** are enantiotopic, the chemical shift difference between these protons is likely to be very small.

Q5. Would you expect the NMR signals of the methyl groups Me^a and Me^b to be equivalent in **77**?

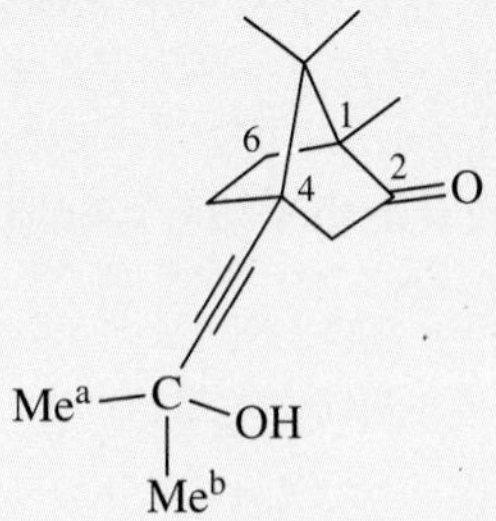

A. The groups Me^a and Me^b in **77** are diastereotopic because of the bicyclic moiety derived from camphor. Chirality on the bicyclic part of the molecule can be viewed in two ways: (1) there are stereogenic centres at both C(4) and C(1); (2) the bicyclic part of the molecule is chiral because the substitution pattern at C(2) (a carbonyl group) differs from that at C(6) (a methylene group). In either case, Me^a and Me^b experience the chirality, induced by the bicyclic part, across the triple bond. The diastereotopism is observed in the ^{13}C NMR spectrum of both **77** and its ethanoate ester. The observed chemical shift differences, though small, are noteworthy because of the large separation of the bicyclic moiety and the 'sensor' nuclei in Me^a and Me^b (see Morris et al.[32]).

Now look up the structure of the amino acid valine. Would you expect the chemical shifts of the protons of the two methyl groups in valine to be the same or different?

Summary of Key Points

- Prochirality has been considered in organic molecules in which the key atoms are either sp^3 or sp^2 hybridized.

- A molecule H_2CAB is prochiral because replacement, separately, of each hydrogen by the same group (usually D for the purpose of definition) gives a pair of enantiomers. The two hydrogens are known as H_R (when replacement by D gives a molecule of R configuration) and H_S (when replacement of the other CH_2 hydrogen results in a molecule of S configuration).

- A carbonyl group in ABC=O is prochiral provided A differs from B; the same holds for the faces of both carbons in an alkene $R^1R^2C=CR^3R^4$, and for both faces of one carbon in $R^1R^2C=CH_2$. The upper face of a carbonyl carbon in ABC=O is given the symbol *Re* if, after using the CIP sequence rules, the priorities, say, O > A > B decrease clockwise when looking at the face from 'above'. In such a case the opposite face of the carbonyl carbon is *Si*.

- In the molecule **19**, for example, the CH_2 hydrogens do not have the same reactivity, and are known as diastereotopic. Similarly, the carbonyl groups in **22** and **23** have diastereotopic faces and, in principle, reagents attack one face preferentially. The selectivity can vary from very small to very large.

- In the enzyme-catalysed reduction of carbonyl groups to alcohols, hydrogen is delivered exclusively to one face. The reverse oxidation of ethanol to ethanal is also completely stereoselective, and involves sole removal of a particular prochiral hydrogen.

- Laboratory attempts to make, say, an enantiomerically pure compound are frequently based on creation of reagents that exhibit high facial selectivity. One cited example was asymmetric hydroboration, which gives alcohols of high enantiomeric purity.

- NMR, the most powerful of the spectroscopies, can be used to identify enantiomer ratios. Firstly, these enantiomers have either to interact or combine with one enantiomer of a suitable compound so that the NMR spectrometer can perceive the resultant diastereomeric complexes or diastereoisomers as separate signals. The suitable compounds can be a chiral solvent, a chiral lanthanide shift reagent or, for an alcohol, an ester made from a chiral acid; in each case the chiral agent must be enantiomerically pure. Integration of observed separate signals for these 'diastereoisomers' permits the enantiomer ratios to be estimated.

- Stereochemical information, both conformational and configurational, can be obtained from the nuclear Overhauser effect (NOE) that operates between two non-bonded protons H^a and H^b in a molecule that are closely separated (< *ca.* 0.4 nm) in space.
- Irradiation of H^a gives an enhanced integral for H^b, and *vice versa*, if the protons are tertiary. With secondary, and especially CH_3 protons, NOEs are less easy to observe.

Problems

8.1. In the structures **78–84**, state, with reasons, whether the hydrogens marked H^a and H^b are homotopic, enantiotopic or diastereotopic.

78

79

80

81

82

83

84

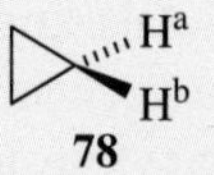

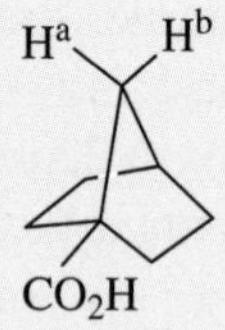

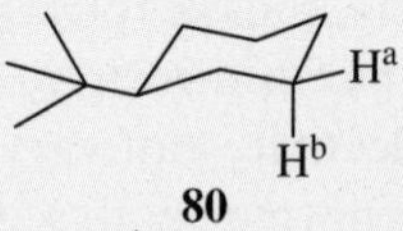

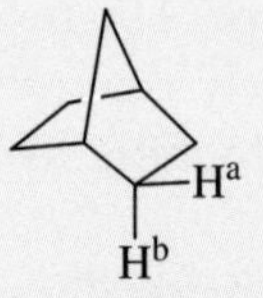

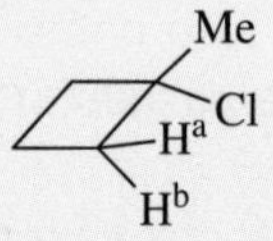

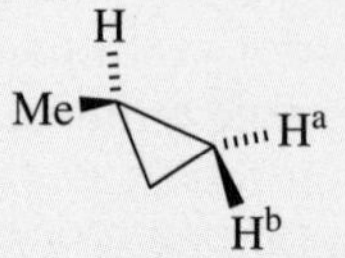

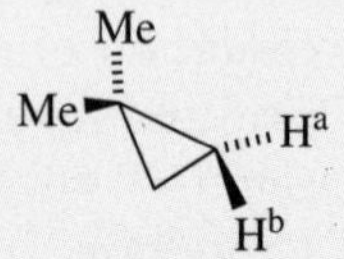

8.2. With reasons, state whether the faces of the carbonyl groups in **85–89** are homotopic, enantiotopic or diastereotopic.

85

86

87

88

89

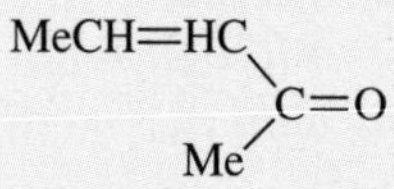

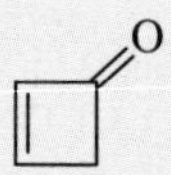

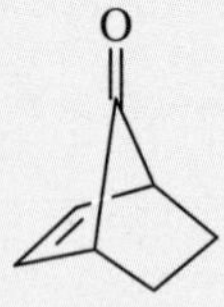

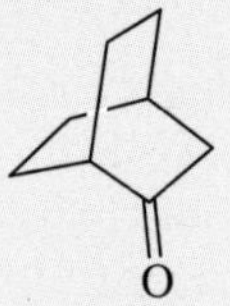

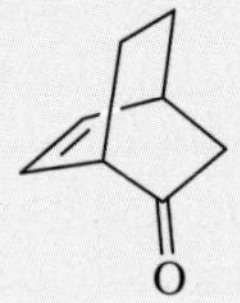

8.3. Assign *Re/Si* configuration to the top face, as drawn, of each sp^2 hybridized carbon and nitrogen in compounds **90–94**.

References

1. K. R. Hanson, *J. Am. Chem. Soc.*, 1966, **88**, 2731.
2. G. Helmchen, in *Houben-Weyl, Methods of Organic Chemistry, Part A, General Aspects*, Vol. E21a, *Stereoselective Synthesis*, Thieme, Stuttgart, 1995, p. 8.
3. E. L. Eliel and S. H. Wilen, *Stereochemistry of Carbon Compounds*, Wiley, New York, 1994, p. 123.
4. R. A. Aitken and S. N. Kilényi (eds.), *Asymmetric Synthesis*, Blackie, London, 1992, p. 27.
5. H. Gunther, *NMR Spectroscopy*, 2nd edn., Wiley, New York, 1980, p. 201.
6. T. Wirth, G. Fragale and M. Spichty, *J. Am. Chem. Soc.*, 1998, **120**, 3376.
7. H. Friebolin, *Basic One and Two Dimensional NMR Spectroscopy*, 3rd edn., VCH, Weinheim, 1998, p. 187.
8. E. W. Garbisch and M. G. Griffith, *J. Am. Chem. Soc.*, 1968, **90**, 6543.
9. M. Karplus, *J. Am. Chem. Soc.*, 1963, **85**, 2870.
10. E. Breitmaier, *Structural Elucidation by NMR in Organic Chemistry*, Wiley, New York, 1993, p. 42.
11. C. A. G. Haasnoot, F. A. A. M. De Leeuw and C. Altona, *Tetrahedron*, 1980, **36**, 2783.
12. W. H. Pirkle, *J. Am. Chem. Soc.*, 1966, **88**, 1837.
13. C. C. Hinckley, *J. Am. Chem. Soc.*, 1969, **91**, 5150.
14. D. Parker, *Chem. Rev.*, 1991, **91**, 1441.
15. P. L. Rinaldi, *Prog. NMR Spectrosc.*, 1983, **15**, 291.
16. J. A. Dale, D. L. Dull and H. S. Mosher, *J. Org. Chem.*, 1969, **34**, 2543.
17. Y. Fukushi, C. Yajima and J. Mizutani, *Tetrahedron Lett.*, 1994, **599**, 9417.
18. H. Takahashi, T. Kusumi, Y. Kan, M. Satake and T. Yasumoto, *Tetrahedron Lett.*, 1996, **37**, 7087.

19. J. M. Seco, S. K. Latypor, E. Quiñoá and R. Riguero, *Tetrahedron*, 1997, **53**, 8541.
20. R. A. Bell and J. K. Saunders, *Can. J. Chem.*, 1970, **48**, 512.
21. R. A. Bell and J. K. Saunders, *Topics Stereochem.*, 1971, **7**, 1.
22. F. A. L. Anet and A. J. R. Bourn, *J. Am. Chem. Soc.*, 1965, **87**, 5250.
23. J. A. Peters, W. M. M. J. Bovée, P. E. J. Peters-van Cranenburgh and H. van Bekkum, *Tetrahedron Lett.*, 1979, 2553.
24. J. A. Peters, J. M. A. Baas, B. van de Graaf, J. M. van der Toorn and H. van Bekkum, *Tetrahedron*, 1978, **34**, 3317.
25. G. Eglinton, J. Martin and W. Parker, *J. Chem. Soc.*, 1965, 1243.
26. J. Stonehouse, P. Adell, J. Keeler and A. J. Shaka, *J. Am. Chem. Soc.*, 1994, **116**, 6037.
27. K. Stott, J. Keeler, Q. N. Van and A. J. Shaka, *J. Magn. Reson.*, 1977, **125**, 302.
28. R. Freeman, *Handbook of Nuclear Magnetic Resonance*, Longman, London, 1988, p. 142.
29. A. Henmann, J. M. Brunel, F. Faure and H. Kolshorn, *Chem. Commun.*, 1996, 1159.
30. D. G. Morris, A. M. Murray, E. B. Mullock, R. M. Plews and J. E. Thorpe, *Tetrahedron Lett.*, 1973, 3179.
31. H. Hamill and M. A. McKervey, *Chem. Commun.*, 1969, 864. *Aust. J. Chem.*, 1982, **35**, 1061.
32. D. G. Morris, A. G. Shepherd, M. F. Walker and R. W. Jemison, *Aust. J. Chem.*, 1982, **35**, 1061.

Further Reading

D. Neuhaus and M. Williamson, *The Nuclear Overhauser Effect*, VCH, Weinheim, 1989.
J. H. Noggle and R. E. Schirmer, *The Nuclear Overhauser Effect*, Academic Press, New York, 1971.

Answers to Problems

Chapter 2

2.1.

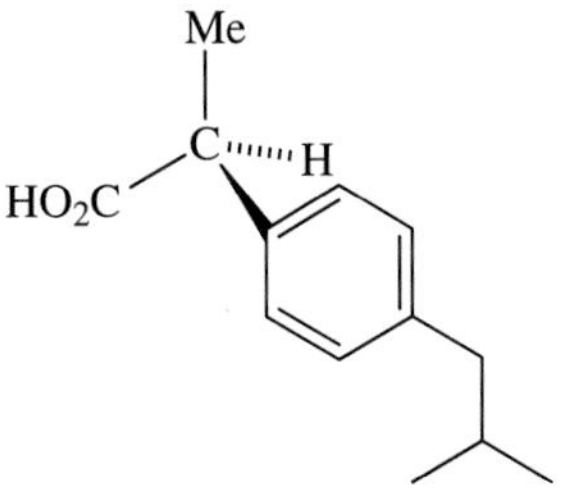

19 (*S*)-ibuprofen

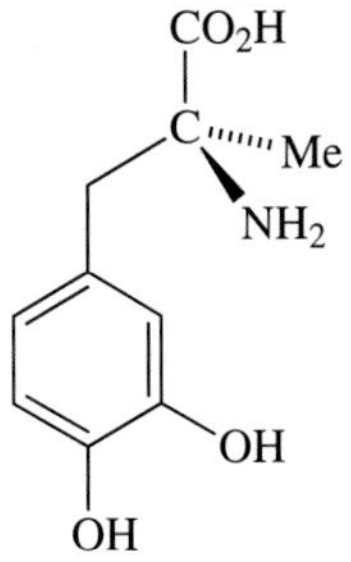

20 (*S*)-methyldopa

2.2.

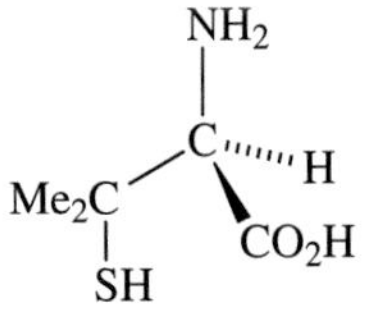

21 (*S*)-penicillamine

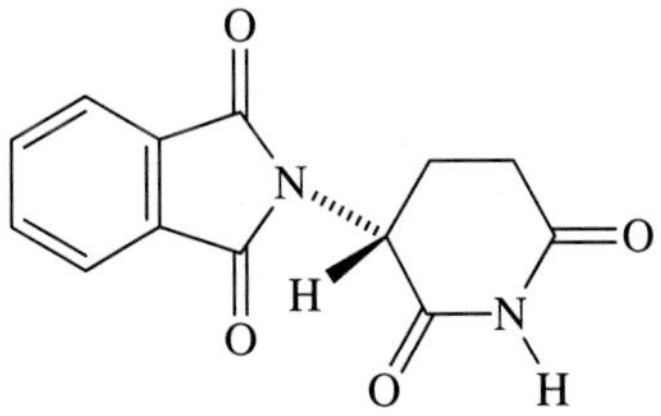

22 (*R*)-thalidomide

2.3.

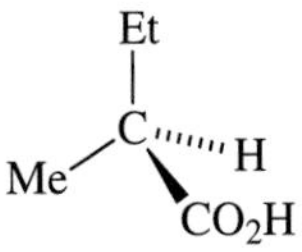

(*S*)-2-methylbutanoic acid

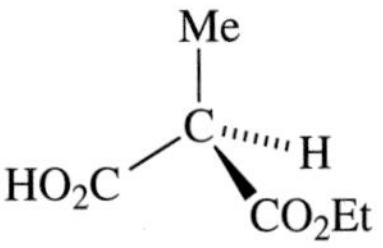

(*R*)-2-(ethoxycarbonyl)propanoic acid

2.4. Compound **23** is *S*; compound **24** is *R*.

2.5.

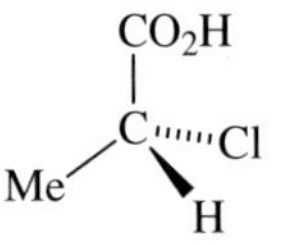

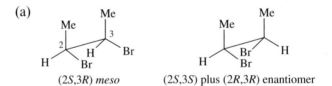

(R)-**25** 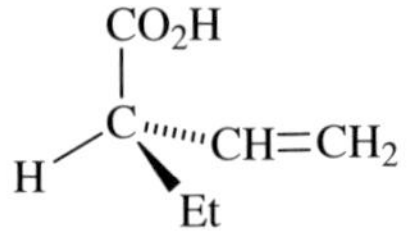   (R)-**26** 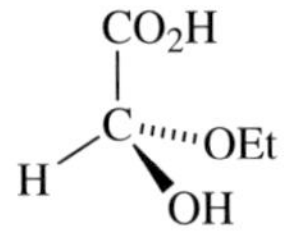   (R)-**27**

2.6. Compounds **28**, **29** and **31** are S; compound **30** is R.

2.7. (i) NO_2 > NO; (ii) Me_3CCH_2 > CH_2=$CHCH_2$; (iii) MeCO > CHO; (iv) Me_3CCH_2 > CH_2Ph > CH=CH_2.

Chapter 3

3.1. A *meso* compound is represented by, for example, (1R,2S), and the identical (1S,2R) structure is not shown. In addition, only one enantiomer of a pair is shown. Where relevant, sawhorse projections are shown in eclipsed conformation, though it should be recognized that this is not the most stable.

(a)

(2S,3R) *meso*   (2S,3S) plus (2R,3R) enantiomer

(b)

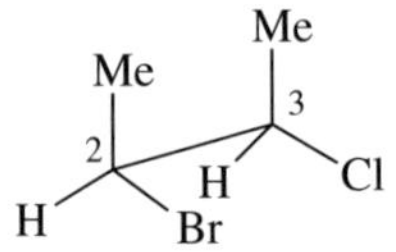

(2S,3R) plus (2R,3S) enantiomer   (2S,3S) plus (2R,3R) enantiomer

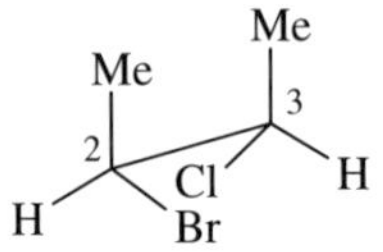

(c)

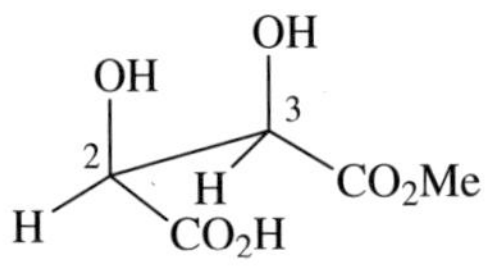

(2S,3R) plus (2R,3S) enantiomer   (2R,3R) plus (2S,3S) enantiomer

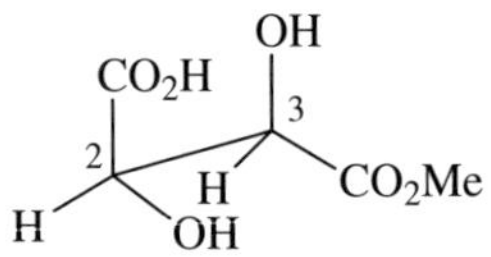

(d)

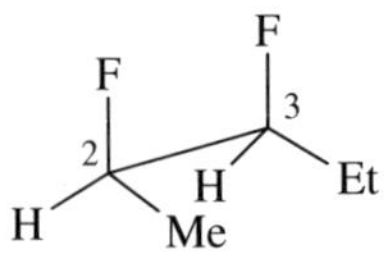

(2R,3S) plus (2S,3R) enantiomer   (2S,3S) plus (2R,3R) enantiomer

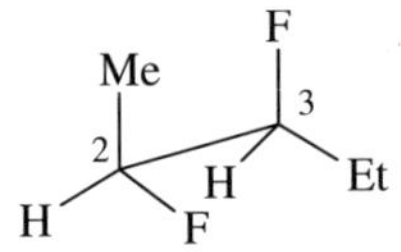

(e)

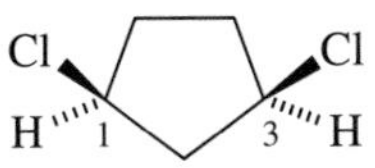

(1S,3R) *meso* (1R,3R) plus (1S,3S) enantiomer

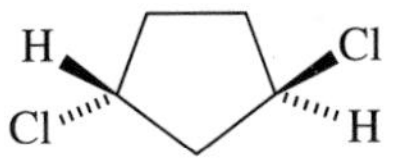

3.2. (a)

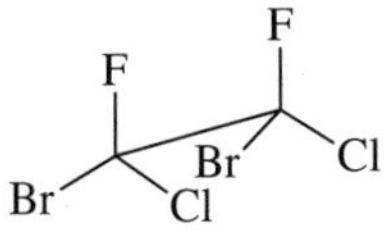

(b)

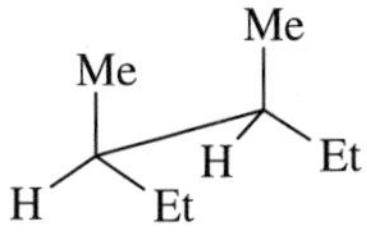

3.3. (a) True. (b) False (a molecule with one stereogenic centre has no diastereoisomers). (c) False (a *meso* compound has two stereogenic centres, but is achiral).

3.4. No. Compound **75** is *meso*.

3.5. Hydrogenation of **76** gives a *meso* product. Hydrogenation of **77** gives a *threo* product.

3.6.

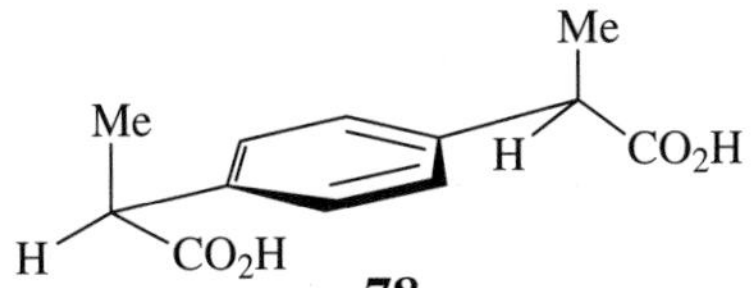

78

3.7.

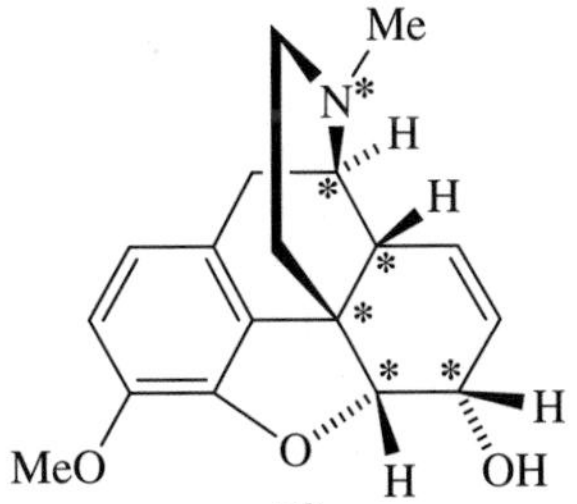

79

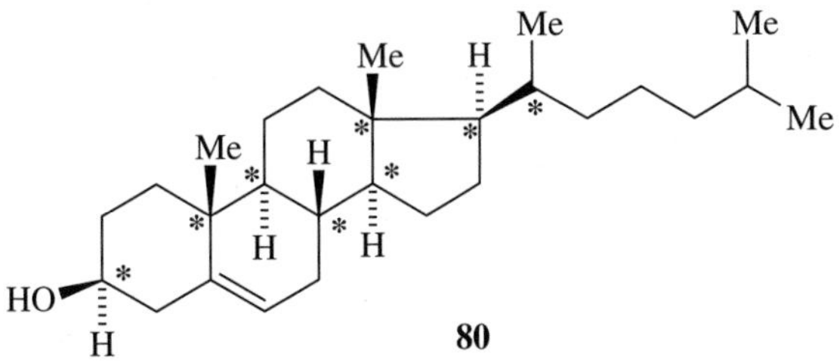

80

3.8. Compounds **81** and **82** are both achiral (they have a plane of symmetry), but compound **83** is chiral.

3.9. Compound **84** has one stereogenic centre and one unsymmetrically disubstituted alkenic double bond; accordingly, **84** can exist as (*R,E*), (*R,Z*), (*S,E*) and (*S,Z*) stereoisomers.

3.10. Truxillic acid exists as five *meso* diastereoisomers. The substituents, as in **85**, can be either up (U) or down (D) (see Mislow[17]):

U U   U U   U D   U D   U U
U U   D D   D U   U U   U D

Chapter 4

4.1. Compound **74** is (*E*); **75** (*Z*); **76** none; **77** (10*E*,12*Z*); **78** (*E*); **79** (*E*); **80** (*E*).

4.2. In nerol (**81**) the configuration is (*Z*); the second double bond has no *E/Z* configuration.

4.3. (a) In **82**, *Z*; in **83**, *E*.

4.4.

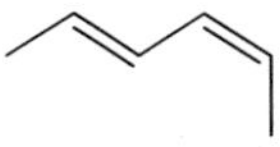

(i) 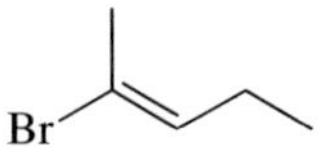　　　(ii) 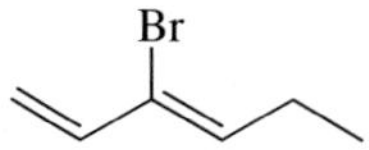　　　(iii) 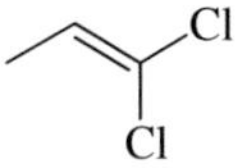　　　(iv)

4.5.

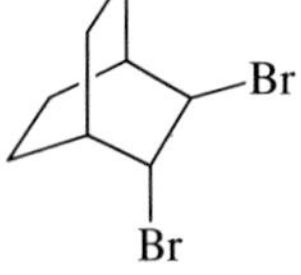

(i) 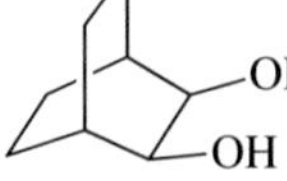　　　(ii) 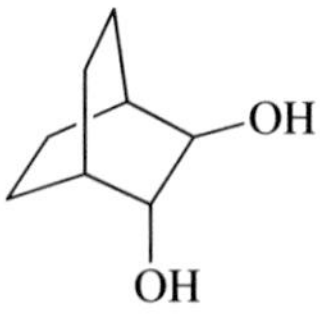　　　(iii) 　　　(iv)

Chapter 5

5.1.

Compound no.	Answer	Explanation
65	Yes	Type $C_{ab}=C=C_{ab}$
66	Yes	Type $C_{ab}-C_{ac}$
67	Yes	Type Ar (*ortho* a,b)–Ar (*ortho* a,c), as in compound **13**
68	Yes	Same type as **8**, but with smaller ring; note $CO_2H \neq CO_2Me$
69	Yes	Same general type as **7**, but smaller ring
70	No	Two exocyclic double bonds means that stereochemistry reverts to alkene type; **70** is (*E*)
71	No	Similar to **70**; **71** is (*E*)
72	No	Two identical ring substituents
73	Yes	Same type as, for example, **7**
74	Yes	A more complicated example than **7**; compound **74** and its minor image are not superimposable
75	Yes	Lone pair on sulfur (not shown) counts as a substituent. Sulfur configurationally stable in sulfoxides
76	Yes	Same type as compound **7**; now the 1,1-dimethylethyl (*t*-butyl) group provides a necessary substituent and also prevents ring inversion.

Chapter 6

6.1. The problems here are best considered with the aid of a molecular model of both enantiomers of each compound. If the enantiomers are not superimposable, the molecule is chiral. (a) Chiral, exists as a pair of enantiomers (*cf. trans*-1,2-dimethylcyclohexane; (b) chiral; (c) achiral (plane of symmetry); (d) chiral; (e) chiral; (f) achiral; (g) chiral; (h) achiral (plane of symmetry).

6.2. The relationship between the CO_2H groups is *trans*, and they cannot approach close enough to react even when the right hand ring of **77**, as drawn, adopts a boat conformation.

6.3. Yes. It is not superimposable on its mirror image.

6.4. The substituents are *cis*, and can approach sufficiently close to each other to form a lactone with loss of water.

6.5. The answer is given in Z. Gore and S. E. Biali, *J. Am. Chem. Soc.*, 1990, **112**, 893.

Chapter 7

7.1. Approach by RS⁻ to the carbonyl carbon of the ester group of **52** is restricted, whereas approach to the methoxy methyl group is open. Reaction to give **53** proceeds *via* a mechanism that is either described as S_N2.

7.2. Displacement of Cl from the anion of **54** can occur *via* the intramolecular concerted pathway shown. The *t*-butyl group ensures that the ring is held in the correct conformation for collinear alignment of the three key atoms. A corresponding collinear alignment is not possible for the anion of **56**, and so no oxirane is formed from this compound.

7.3. Compound **57** reacts with NaCN to produce the cyanohydrin ion shown, and its epimer, in a reaction that is probably reversible. The anion is converted into **58** in an intramolecular exocyclic nucleophilic displacement, as shown. The 'wrong' epimer cannot react to give **58**, but if the initial reaction is reversible, the 'right' epimer (below) will be formed and this will give **58**.

Chapter 8

8.1. In **78–84** the answer is arrived at by consideration of the compounds obtained when H^a = D, H^b = H, and H^a = H, H^b = D. In homotopic cases the same compound is obtained, in enantiotopic cases enantiomers are obtained, and in diastereotopic cases, diastereoisomers result. In the diastereotopic cases there are always at least two stereogenic centres in the molecule after a monodeuteriation. Thus H^a and H^b are homotopic in **78**; enantiotopic in **79**; diastereotopic in **80** [note that the *t*-butyl group prevents ring inversion; in the deuteriated molecule, C(1) is a stereogenic centre]; diastereotopic in **81** (in the deuteriated molecule, both bridgehead carbons are stereogenic); diastereotopic in **82**; diastereotopic in **83**; and enantiotopic in **84**. Note that the three cyclopropane derivatives, **78**, **83** and **84**, are homotopic, diastereotopic and enantiotopic, respectively.

8.2. The answers here can be obtained by comparison of the products formed as the result of attack at the two faces of the carbonyl group by a reagent, *e.g.* $LiAlH_4$ or MeMgI, as appropriate, chosen to give, where possible, a stereogenic centre after reaction at the carbonyl carbon. Homotopic carbons give identical products, enantiotopic carbons give enantiomers and diastereotopic carbons give diastereoisomers. The faces are enantiotopic in **85**; enantiotopic in **86**; diastereotopic in **87** [the products of reaction of **87** with, *e.g.* $LiAlH_4$, are diastereoisomeric; draw these diastereoisomers and note that both are racemic (as is the case for, *e.g. cis*- and *trans*-1,4-dichlorocyclohexane]; enantiotopic in **88** (this may be more apparent by turning the molecule anti-clockwise by 120° about an axis through the bridgehead carbons); and diastereotopic in **89**.

8.3. **90**, *Si*, *Re*; **91**, *Si*; **92**, *Re*, *Si*; **93**, neither *Si* nor *Re*; **94**, *Si*, *Re*.

Subject Index